30-SECOND
PHYSICS

30-SECOND
PHYSICS

The 50 most fundamental
concepts in physics, each
explained in half a minute

Consultant Editor
Brian Clegg

Contributors
Philip Ball
Brian Clegg
Leon Clifford
Frank Close
Rhodri Evans
Andrew May

Illustrations
Steve Rawlings

Ivy Press

First published in the UK in 2016 by
Ivy Press
210 High Street, Lewes,
East Sussex BN7 2NS, UK
www.ivypress.co.uk

British Library Cataloguing-in-
Publication Data
A CIP catalogue record for this
book is available from the
British Library.

ISBN: 978-1-78240-312-8

This book was conceived,
designed and produced by
Ivy Press

Creative Director Michael Whitehead
Publisher Susan Kelly
Editorial Director Tom Kitch
Art Director James Lawrence
Project Editor Jamie Pumfrey
Editor Charles Phillips
Designer Ginny Zeal
Glossaries Text Brian Clegg

Typeset in Section

Printed and bound in China

10 9 8 7 6 5 4 3 2 1

CONTENTS

INTRODUCTION
Brian Clegg

Physics is arguably the ultimate science, which describes how everything works. There was more than an element of truth in Ernest Rutherford's witty statement: 'All science is either physics or stamp collecting.' In his day, the other sciences primarily concerned themselves with collecting information and structuring it, rather than looking for explanations. That's far less true now, but nonetheless, physics remains at the heart of scientific discovery.

The word 'physics' comes from the Latin *physica*, which meant natural science in general ('science' covered all knowledge) – reflecting the way that the term had been used by Greek philosopher Aristotle. But from the 18th century onwards, physics became more tightly defined as the science of non-living matter and energy, with the arbitrary restriction of not including chemical elements, compounds and their reactions. Hence it ranged from mechanics and light to gravity, the nature of matter, astronomy and cosmology.

Now physics includes everything from the extremely small (such as the nature of subatomic particles) to the mechanisms responsible for the formation of the universe. While the aim of physics is to explain the workings of the physical world, it also results in huge practical developments. Technology explicitly using quantum physics, for instance, produces around thirty-five per cent of GDP in developed countries, and our exploration of light has brought everything from X-rays to Wi-Fi.

It's easy to get turned off physics at school, because some basic physics, like mechanics and optics, can seem tedious. But physics provides us with the most mind-bending aspects of science. Whether we're dealing with quantum theory or relativity, physics makes concepts like black holes, time travel and teleportation real.

Our exploration of physics begins with stuff – matter – with atoms at its heart. As well as the familiar forms of solid, liquid and gas, we explore plasmas and the mysterious world of antimatter. But matter would not get us far without light: this is our second topic. We tend to think of light

as the stuff we see with, but it is much more. The full electromagnetic spectrum – light being a self-supporting interaction between electricity and magnetism – ranges from radio, through microwaves, infrared, the visible spectrum, ultraviolet, X-rays and gamma rays. The part we see is a tiny segment of the whole.

Light inevitably brings in colour, and light's interactions with matter such as reflection and refraction. We now frequently model light as a collection of quantum particles, or a disturbance in a quantum field, which leads us neatly to the next section on quantum theory. Exploring concepts such as wave/particle duality, the uncertainty principle and entanglement, we begin to experience the strangeness of the behaviour of light and matter on this scale.

Electromagnetism, responsible both for light and most of the mechanical interactions of matter, is part of our next section on forces. As well as the four fundamental forces of nature, we consider orbits and how best to describe the action of forces. Generally speaking, forces produce the subject of our next section – motion. Here Newton's laws jostle alongside Einstein's special relativity and its combination of space and time in a single entity, spacetime.

To make motion or pretty well anything else happen, we need energy, the topic of the penultimate chapter. Energy is at the heart of everything from living things to machines. A particular subset of machines – steam engines – gave birth to our final topic, thermodynamics. Originally devised to improve steam technology, the laws of thermodynamics tell us much more – including the potential fate of the universe.

Whatever our interest, physics is there, helping us understand the world around us.

"

For those who want some proof that physicists are human, the proof is in the idiocy of all the different units which they use for measuring energy

RICHARD FEYNMAN

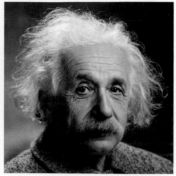

As far as the laws of mathematics refer to reality, they are not certain; and as far as they are certain, they do not refer to reality

ALBERT EINSTEIN

All science is either physics or stamp collecting

ERNEST RUTHERFORD

If I could remember the names of all those particles I'd be a botanist

ENRICO FERMI

It should be remarked, to begin with, that we have no right to assume that any physical law exists, or if they have existed up to now, that they will continue to exist in a similar manner in the future

MAX PLANCK

The future of chemistry rests and must rest, with physics

C. P. SNOW

To understand hydrogen is to understand all of physics

VICTOR WEISSKOPF

My Design in this Book is not to explain the Properties of Light by Hypothesis, but to propose and prove them by Reason and Experiments

ISAAC NEWTON

Why, sir, there is every possibility that you will soon be able to tax it!

MICHAEL FARADAY

Allegedly said to Gladstone when he asked about the practical worth of electricity

It is wrong to think that the task of physics is to find out how nature is. Physics concerns what we can say about nature

NIELS BOHR

MATTER

amorphous solid A solid in which atoms or molecules are not arranged in a repeating crystalline structure, but are scattered in a less structured fashion. The best-known amorphous solid is glass, although plenty of other materials – from plastics to some kinds of metal – can be amorphous.

electromagnetic field A model of the way electricity and magnetism act. The field can be thought of as being like a physical contour map. In modern physics, the field is 'quantized' – made up of distinct parts, where a change in the field can be represented as a particle called a photon.

electron A fundamental subatomic quantum particle with a negative electrical charge. Electrons occupy fuzzy 'orbitals' in the outer parts of atoms, jumping from one orbit to another as a result of absorbing or giving off a photon of light. Electrons carry the charge when an electric current flows.

gravitational mass The property of matter that makes it attract other matter. The greater the gravitational mass, the greater the force with which a body will attract another body. Identical in size to inertial mass.

inertial mass The property of matter that makes it difficult to change its state of motion. The more inertial mass it has, the more force it takes to start it moving or to slow it down when it is moving. Inertial mass is identical in size to gravitational mass.

neutrino An uncharged fundamental quantum particle with very low mass, produced during nuclear reactions. The neutrino was predicted in 1930 to explain the loss of energy during a nuclear reaction, but it was not detected until 1956 because it has very little interaction with matter. The name means 'little neutral one'.

neutron A neutral or uncharged quantum particle, most frequently found in the nucleus of an atom, composed of three fundamental particles: one up quark and two down quarks. Atoms of the same element can have differing numbers of neutrons in their nucleus; such variants are called isotopes. For instance, hydrogen, the most basic atom, usually has one proton and no neutrons in its nucleus, but it also comes in the form known as deuterium, with one proton and one neutron in the nucleus.

Newton's second law of motion Newton's second law originally took the form that a change of motion is proportional to the force applied and takes place in the direction of the application of force, but it is now simply stated as $F=ma$, where F is the force applied, m is the mass of the object the force is applied to and a is the resultant acceleration – the rate of change of the object's velocity.

photon A massless quantum particle of light. Light can be described as a wave, a particle or a disturbance in an electromagnetic field. These are all models that help us understand it – light itself is just light. Describing light as a particle is helpful when dealing with the interaction between light and matter, and it became essential when Einstein described the way energetic photons knock electrons out of metals, producing an electric current. The energy of a photon is equivalent to the light's colour. The photon is the carrier particle of the electromagnetic force: when two objects interact electrically or magnetically, photons travelling between the objects carry the force.

proton A positively charged quantum particle, most frequently found in the nucleus of an atom, composed of three fundamental particles: two up quarks and one down quark. The number of protons in an atom determine which element the atom is – the 'atomic number' of an element is the number of protons it has. A single proton makes up the nucleus of the most basic atom, hydrogen.

quark A fundamental quantum particle, with either two-thirds the charge of a proton or one-third the charge of an electron. Quarks come in six 'flavours': up, down, charm, strange, top and bottom. Triplets of quarks make up protons and neutrons, while quark-antiquark pairings make up mesons.

ATOMS

the 30-second theory

It is marvellously convenient that all ordinary matter is made up of atoms, for that greatly simplifies the job of explaining its properties. A great deal, from the shapes of crystals to the stretchiness of rubber, can be explained on the basis of how atoms join and stack together in groups. The dizzying diversity of behaviour amongst organic (carbon-based) substances, from drugs to solvents to DNA, originates in the unions of just a few types of atom into molecules with different shapes and physical and chemical properties. Indeed, the entire physical world can be described with only 92 or so basic building blocks, only a few dozen of which are particularly common. Of course, that isn't all there is to it: atoms are misnamed, for they are not really indivisible (the meaning of the Greek *a-tomos*, from which they got their name). But atoms are the fundamental unit of chemical theory: each element comprises atoms with the same number of nuclear protons and orbiting electrons (the number of neutrons may vary in different isotopes) and it is primarily the disposition of electrons that determines how atoms react chemically with others. Thanks to the atomic granularity of matter, everything from the hardness of diamond to the toxicity of lead can be understood using the same theoretical framework: the quantum theory of atoms.

3-SECOND THRASH
Every material thing in our familiar world is made of atoms, which are the fundamental units of chemical theory.

3-MINUTE THOUGHT
Atoms are small enough to show quantum-mechanical behaviour, such as exhibiting wave-like properties under the right conditions. Interference of atom waves has been observed many times in recent decades, even for single atoms. Wave interference effects have also been observed for molecules containing more than 100 atoms each. Whether there is any fundamental size limit on such phenomena is a question still under discussion.

RELATED TOPICS
See also
MASS
page 18

SOLIDS
page 20

LIQUIDS
page 22

3-SECOND BIOGRAPHIES
DEMOCRITUS
C.460–C.370 BCE
Greek philosopher who proposed that matter was made up of atoms

JOHN DALTON
1766–1844
English chemist who formulated the basics of modern atomic theory

JEAN PERRIN
1870–1942
French physicist who helped confirm the reality of atoms

30-SECOND TEXT
Philip Ball

Atoms are building blocks – from the galaxies in space to the Earth itself to the smallest piece of matter.

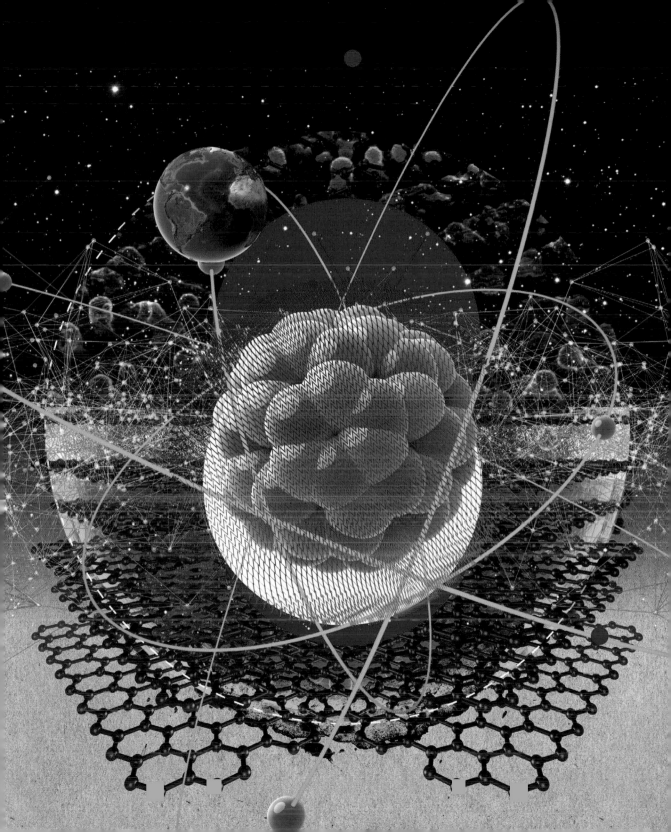

MASS

the 30-second theory

In the basement of a building on the outskirts of Paris is a lump of metal (90 per cent platinum and 10 per cent iridium) in a climate-controlled safe; this lump of metal defines 1 kg (approximately 2 lb 3 oz) of mass. But what is mass? From Newton's second law of motion, mass is the property of a body that determines how much force is necessary to accelerate it. Alternatively, mass determines the strength of gravity between two objects a given distance apart. The first of these definitions is of 'inertial mass' and the second of 'gravitational mass'; Einstein showed in his principle of equivalence that they are the same. However, mass and weight are different; we might say 'I weigh 76 kg (12 stone)', but that would be our mass, not our weight. Our weight would change if we were on the Moon, but our mass would be unchanged. In space, with no gravity around, if you want to accelerate a massive object you need a larger force than if you want to accelerate a less massive object. The only particles we know of with zero mass are some bosons like photons and gluons. A neutrino has very close to zero (but not quite zero) mass; it is the next least massive particle.

RELATED TOPICS
See also
GALILEO
page 26

PHOTONS
page 40

FORCE & ACCELERATION
page 78

3-SECOND BIOGRAPHIES
GALILEO GALILEI
1564–1642
Italian natural philosopher who experimented on the motion and acceleration of bodies

ALBERT EINSTEIN
1879–1955
German-born physicist whose special and general relativity gave a new understanding of mass

30-SECOND TEXT
Rhodri Evans

3-SECOND THRASH
Mass, measured in kilograms (or pounds), determines how hard it is to accelerate an object, and how strong the force of gravity is between it and, for example, the Earth.

3-MINUTE THOUGHT
Einstein showed in his special theory of relativity that mass and energy are related through probably the most famous equation in physics, $E=mc^2$ (E is energy and c is the speed of light). We can therefore think of mass as concentrated energy; in nuclear power plants and in the Sun mass is converted to energy, but when we burn oil, for example, the energy comes from changes in chemical bonds.

While your mass is the same on the Moon as on Earth, your weight differs.

SOLIDS

the 30-second theory

Solids are generally the densest state of ordinary matter, consisting of atoms packed closely together and held in place by chemical bonds. It is hard to generalize about their properties except in negative terms: they tend not to flow (like liquids) and will not expand (like gases) to fill available space. Solids are often strong and offer resistance to forces that might distort their shape: rocks, metals and ceramics are the archetypal examples. Beyond this, they may show a wide range of properties. Their atoms might be stacked in orderly, repeating arrangements, making them crystals, or they might be disorderly, as in amorphous solids such as glasses. Some solids are soft and elastic, because their molecular constituents are only weakly bonded to each other and can store energy when displaced. Others are rigid and perhaps prone to brittle failure. Some conduct electricity because they contain mobile electrons; others are insulating because the atoms retain all their electrons tightly. There is no rigorous definition of a solid. Some gels hold their shape even though they are mostly liquid trapped in a network of polymer strands. Aerogels may be 99 per cent empty space, and silica aerogel in a vacuum can have a lower density than ambient air. Some substances with apparently solid-like resistance to deformation, such as bitumen, may flow sluggishly.

RELATED TOPICS
See also
ATOMS
page 16

LIQUIDS
page 22

GASES
page 24

3-SECOND THRASH
Solids are usually dense and can resist being deformed by squashing or pulling.

3-MINUTE THOUGHT
Neutron stars may host the ultimate solids. Their outer crust should be a super-dense crystalline lattice of atomic nuclei, perhaps of iron, in a sea of electrons. Some atomic nuclei persist in the inner crust, rich in neutrons formed from crushed protons and electrons. A matchbox of this solid would weigh around 5 billion tonnes. In the inner core no nuclei survive: whatever weird stuff it contains, the concept of solid has no clear meaning at such densities.

3-SECOND BIOGRAPHIES
NEVILL FRANCIS MOTT
1905–96
English physicist who studied the electronic properties of solids

FREDERICK CHARLES FRANK
1911–98
English physicist who advanced the theory of crystal structures

NEIL ASHCROFT
1938–
English physicist who specializes in the structure of solids at high pressures

30-SECOND TEXT
Philip Ball

Generally, solids contain atoms tightly bound in close formations – so will not easily deform or flow.

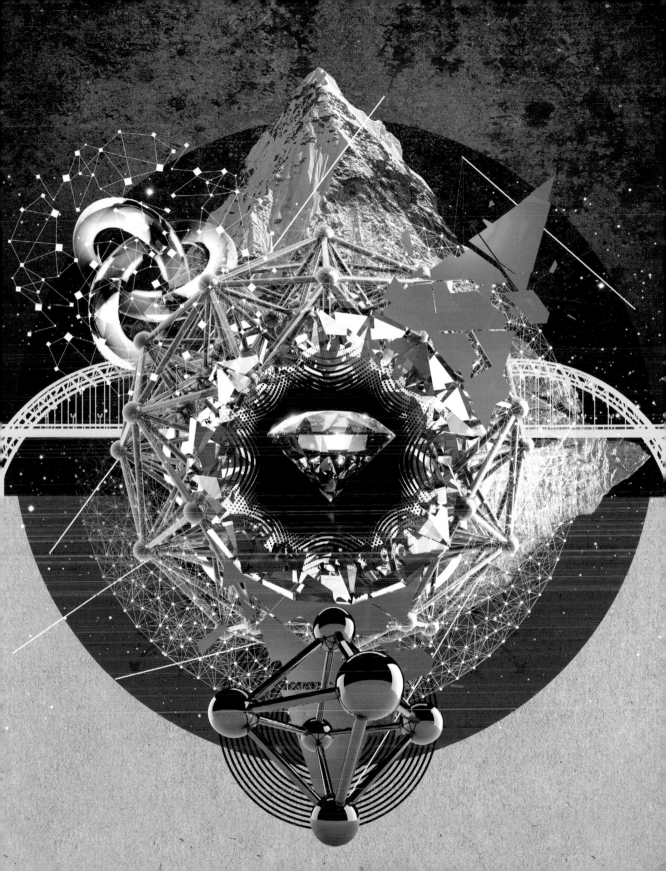

LIQUIDS

the 30-second theory

If a bunch of atoms get cold enough, they'll solidify; if hot enough, they'll vaporize. So solids and gases are a given – but the third common state of matter, liquid, is a curious intermediate, neither totally orderly like a crystalline solid nor totally disordered like a gas. Attractive forces between atoms bind the particles into a dense mass, but they remain mobile, giving liquids fluidity and disorganized structures. Over distances of a few molecular diameters liquids have some regularity simply due to the constraints of packing the particles together. In some liquids such as water, where there are weak chemical bonds between molecules with a certain geometrical arrangement, this short-ranged order is even more pronounced. But over longer ranges any regularity is lost. Because they are poised between order and disorder, liquids are a challenging state of matter to understand and describe, and liquid-state theory is still an evolving field. One particular complication is that the molecular motions are not independent, as in a gas, but correlated: the movement of one molecule affects that of others nearby. That needs to be accommodated into explanations of, for example, liquid viscosity and flow. Liquids are closely related to glasses, where the molecules have become so slow as to be almost immobile, frozen into disorder.

RELATED TOPICS
See also
ATOMS
page 16

SOLIDS
page 20

3-SECOND THRASH
Liquids are a complex intermediate state between the perfect order (in principle) of solids and the perfect disorder of gases.

3-MINUTE THOUGHT
In liquid helium, which persists at temperatures close to absolute zero, quantum-mechanical effects do strange things. The helium atoms can all enter the same quantum state, which means that they behave like a single, giant collective particle. This in turn means that they can flow without any viscous resistance, allowing them to creep up the sides of a vessel and over the rim. Such behaviour is called superfluidity.

3-SECOND BIOGRAPHIES
JOHANNES DIDERIK
VAN DER WAALS
1837–1923
Dutch physicist who established the theory of liquids and how they relate to gasses

JOHN GAMBLE KIRKWOOD
1907–59
American physicist who used the forces between molecules to statistically model liquids

PIERRE-GILLES DE GENNES
1932–2007
French Nobel laureate who investigated the way liquids spread on and wet surfaces

30-SECOND TEXT
Philip Ball

Between order and disorder. Liquids flow because their atoms, while bound together by attractive forces, are still mobile.

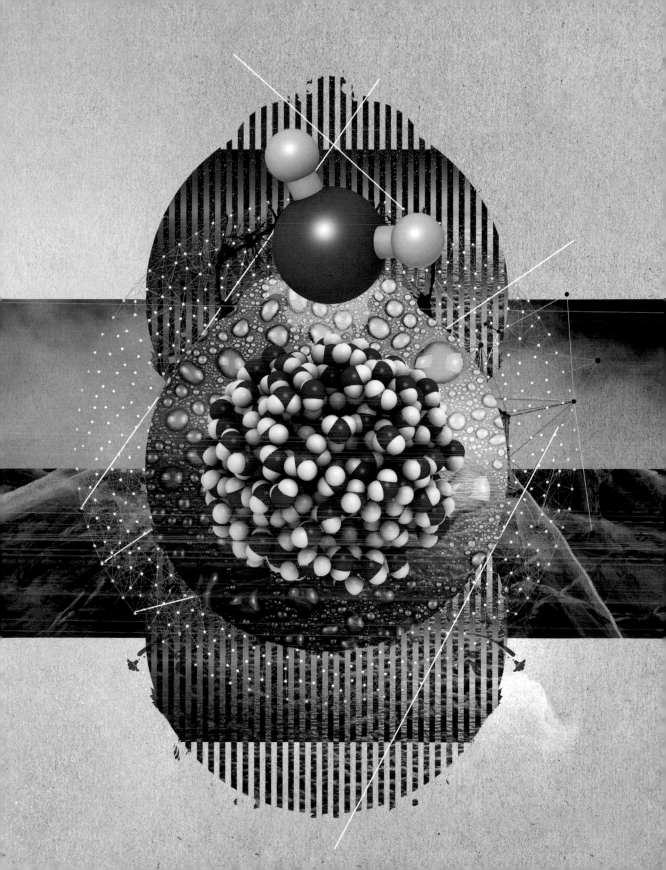

GASES

the 30-second theory

Like a liquid, a gas is a fluid state of matter, but because the particles (atoms or molecules) that make it up are moving around much faster than those in a liquid, with greater energy, their mutual attraction has little effect on the gas's behaviour. The result is that the gas doesn't form a surface, and expands to fill the space available. When the gas particles reach a barrier they collide with it, producing a force on the barrier, which is felt as the pressure of the gas. Reduce the size of the container and the particles have less far to travel, so collide with the walls more frequently. The result is that the pressure times the volume of the gas remains constant, a relationship known as Boyle's law. The pressure is also increased by pushing up the temperature, meaning that the particles move faster – this is called both Amontons' law and Gay-Lussac's law. The result of these two observations is that at constant pressure, the volume of gas has to increase and decrease with temperature, known as Charles' law. These three laws are combined to form the gas law: the pressure times the volume divided by the temperature of a gas remains constant.

RELATED TOPICS
See also
ATOMS
page 16

LIQUIDS
page 22

3-SECOND THRASH
Gas atoms or molecules are moving too quickly to have much attraction, so they fill the space available and obey simple statistical 'laws' linking temperature, pressure and volume.

3-MINUTE THOUGHT
We could not sensibly deal with gases without statistics, as there are far too many atoms or molecules in a body of gas to work out individual motion. Measurements like temperature and pressure are statistical, combining the effects of billions of gas molecules. At room temperature, air molecules are flying around at around 500 metres per second (1640 feet per second), but because their mass is so small, the kinetic energy of each molecule is only around 6×10^{-21} joules – negligible without the combination of many, many molecules.

3-SECOND BIOGRAPHIES
ROBERT BOYLE
1627–91
Anglo-Irish natural philosopher

JACQUES-ALEXANDRE-CÉSAR CHARLES
1746–1823
French scientist after whom the relationship between gas temperature and volume was named

JOSEPH-LOUIS GAY-LUSSAC
1778–1850
French chemist and physicist who worked widely on gases

30-SECOND TEXT
Brian Clegg

Gas atoms' and molecules' speed of movement overpowers the attraction between them – so gasses tend to expand.

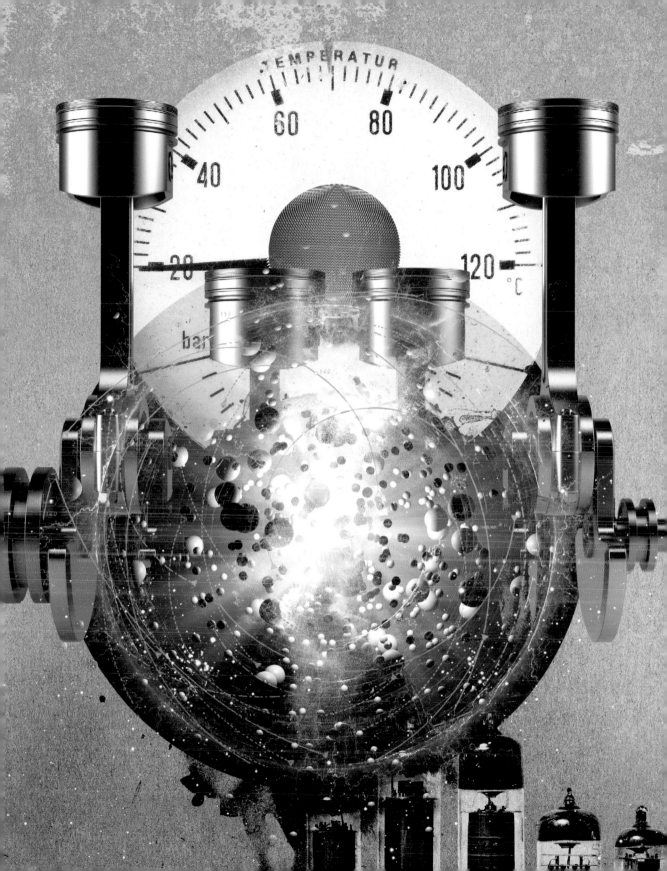

15 February 1564
Born in Pisa

1581
Enters University of Pisa to study medicine

1583
Switches to studying mathematics

1585
Leaves the University of Pisa without graduating

1589
Appointed professor of geometry at the University of Pisa

1592
Let go by Pisa, appointed professor at the University of Padua

1591–1604
Does important work on mechanics, falling bodies and acceleration

1609
After hearing of a telescope, Galileo builds his own

1610
Observes the Moon, discovers Jupiter's four large moons and sees phases of Venus

1610
Leaves the University of Padua

1610
Publishes his early telescope findings in *The Starry Messenger*

1616
Formally cautioned by the Roman Catholic Church for promoting the 'heliocentric' model

1623
Publishes *The Assayer*

1632
Publishes *Dialogue Concerning the Two Chief Systems of the World*

1633
Guilty of violating the terms of his 1616 caution, and sentenced to imprisonment, commuted to house arrest

1638
Draws together his life's work in *Two New Sciences*

8 January 1642
Dies in Florence

GALILEO

If Newton is the father of physics, Galileo can be thought of as the grandfather. Born in Pisa in 1564, he was the son of a lute player and musical theorist, Vincenzo, who did important work on the relationship between the tension, mass and cross-sectional area of a string and the note that it would play. Galileo's uncle was a doctor and this is the career that Vincenzo wanted for his son, but after two years of medical studies Galileo persuaded his father to let him switch to mathematics. After four years at the University of Pisa he left without graduating. This was not unusual for Italians of Galileo's social class at this time. He then spent four years tutoring mathematics and broadening his education to include literature such as Dante's *The Inferno*, and in 1589 he was appointed a professor at Pisa, where he had studied four years before.

This position lasted only three years, partly due to Galileo's increasingly vocal opposition to Aristotelean philosophy – Galileo was one of a new breed of scientists who questioned the teachings of the Greek philosopher and argued that experimentation was the way to get at the true nature of the world. However, through some influential friends he got a professorship in 1592 at the more prestigious University of Padua, where he would stay until resigning in 1610. During his 18 years in Padua he did important work on the motions of bodies, laying the foundations for Newton's three laws of motion.

In 1609 his life took an abrupt turn when he heard of the invention of the telescope and decided from the description to build his own. In late 1609 and throughout 1610 he made observations that showed the Sun and planets could not all be orbiting the Earth, and he became a vociferous supporter of Copernicus' 'heliocentric' (Sun-centred) model. By 1616 this had got him into trouble with the Roman Catholic Church, which expressly forbade him to continue supporting 'Copernican astronomy'. Unable to remain neutral on the subject, in 1632 he published *Dialogue Concerning the Two Chief Systems of the World*, which the Church decided did not present a balanced argument between the two models. In 1633 the Church decreed that Galileo had breached the terms of his 1616 warning, and they sentenced him to imprisonment for heresy. This was commuted to a more lenient sentence, but for the rest of his life Galileo was under house arrest. In 1638 he published *Two New Sciences*, summarizing his life's work, and four years later on 8 January 1642 he died peacefully in Florence.

Rhodri Evans

PLASMA

the 30-second theory

When a gas is heated to

extreme temperatures or subjected to strong electromagnetic fields, it changes phase and becomes a plasma – the fourth state of matter after solid, liquid and gas. The Sun is made from plasma. When plasmas form, molecular bonds between atoms break and electrons separate from parent atoms. An atom left positively charged when it is dissociated from a negatively charged electron is known as an ion: it is this ionization that distinguishes a plasma from a gas. A gas is not ionized, consisting of freely moving atoms or molecules that are not disassociated from their respective electrons and so remain individually electrically neutral. A plasma consists of electrically charged ions and electrons, although the plasma as a whole is usually electrically neutral. If an electromagnetic field is applied to a plasma, the positively charged ions and negatively charged electrons will move in opposite directions, creating an electrical current. All plasmas can conduct electricity and every material becomes electrically conductive when it is in a plasma state. This means that, unlike gases, plasmas can be confined without the use of solid walls by applying electromagnetic fields – and that plasmas within electromagnetic fields can exhibit shape and structure rather than diffuse away like gases.

RELATED TOPICS
See also
SOLIDS
page 20

LIQUIDS
page 22

GASES
page 24

3-SECOND THRASH
Plasma is the strangest of the four fundamental states of matter: it consists of a soup of electrically charged atoms, or ions, and their dissociated electrons.

3-MINUTE THOUGHT
Solids, liquids and gases make up most of the world around us, but it is plasma that is the most abundant state of normal matter in the universe. Stars are made from plasma and a thin plasma occupies the huge gulfs of space between the galaxies. For 380,000 years following the first second of creation, after normal matter and the forces of Nature had condensed from the big bang, the entire universe consisted of plasma.

3-SECOND BIOGRAPHIES
IRVING LANGMUIR
1881–1957
American chemist who coined the term plasma for ionized gases

HANNES ALFVÉN
1908–95
Swedish electrical engineer who argued that a plasma is an electrically conducting fluid

JAMES VAN ALLEN
1914–2006
American physicist who discovered that the Earth is encircled by plasma

30-SECOND TEXT
Leon Clifford

Plasma, plasma everywhere ... The Sun and stars – and even space – consist of plasma.

ANTIMATTER

the 30-second theory

Every particle has an antiparticle
with the same mass but opposite values of
properties like electrical charge. The antiparticle
of the negatively charged electron is the
positively charged positron, while that of the
proton is the negatively charged antiproton.
Even a neutron has its antiparticle with opposing
properties like magnetic moment. When particle
and antiparticle meet they can mutually
annihilate, their mass converted into energy;
this is $E=mc^2$ at work. In particle physics, mutual
annihilation is used in colliders of electrons and
positrons, or protons and antiprotons. It is also
used in medicine in PET (positron emission
tomography) scanners. Conversely, energy
can materialize in counterbalanced matter and
antimatter, as happened during the big bang.
The fundamental laws show no preference for
matter over antimatter, which makes the
apparent dominance of matter over antimatter
in the universe an unresolved mystery. A positron
and an antiproton make an atom of antihydrogen.
Anti-nuclei and anti-elements can exist in
principle, but nothing beyond anti-hydrogen
has yet been made. It is not possible to use
antimatter to solve the world's energy problems,
nor to make bombs. All antimatter first has to be
made, which expends energy. To make a gram of
antimatter would take billions of years.

RELATED TOPICS
See also
ATOMS
page 16

QUANTA
page 58

QED
page 68

3-SECOND BIOGRAPHIES
PAUL DIRAC
1902–84
English physicist who predicted
the existence of antimatter

CARL ANDERSON
1905–91
American physicist who
discovered the positron in
cosmic rays in 1932

30-SECOND TEXT
Frank Close

*Theoretically, the
Big Bang produced
matching quantities
of matter and
antimatter – so why
can't physicists detect
equal amounts?*

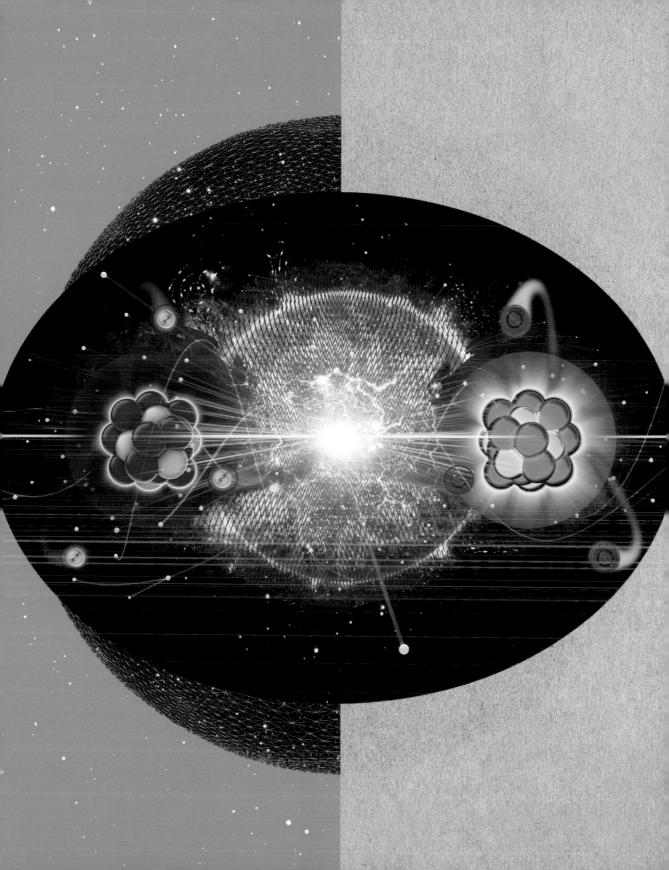

LIGHT ◑

birefringent For most transparent materials, a single 'refractive index' determines how much light bends when it travels from air into the material (and out again). When a material is birefringent, its refractive index changes depending on how the light is polarized. The outcome is that unpolarized light splits into two, producing two images of something seen through the material. One of the best known such materials is Iceland spar.

cosmic microwave background radiation The universe is thought to have first become transparent around 300,000 years after it came into existence. The light that started to travel across the universe then is still detectable. At the time it was high-energy gamma rays, but as the universe has expanded this has been red-shifted to microwaves, which are detectable in all directions in the sky, forming the cosmic microwave background.

electromagnetic waves Light is an interaction of electricity and magnetism, which can be described as a wave, a particle or a disturbance in a field. The earliest well-developed description was of light as a wave, consisting of an electrical and magnetic wave at right angles to each other. This is the same for all kinds of light from radio to gamma waves, not just visible light – the only distinction between different kinds is the wavelength (or frequency) of the wave.

gamma rays High-energy electromagnetic radiation (light). Gamma rays are produced by nuclear reactions and have wavelengths less than 10 picometres (1/100th of a nanometre).

Iceland spar A form of transparent calcite (calcium carbonate) crystal that is birefringent, bending light differently depending on its polarization and so producing two images of something seen through it.

photoelectric effect Some materials produce an electric current when exposed to light. This current is caused by photons of light adding energy to electrons, boosting them away from the atoms in the material to be free to move and carry a current. To explain the photoelectric effect, which is dependent on the frequency of the light but not its intensity, Einstein suggested that light was made up of photons rather than a continuous wave. This is the work for which he won his Nobel Prize for physics.

polarized light The wave model describes light as a side-to-side wave as the light moves along, with an electrical wave at right angles to a magnetic wave. The direction of the electrical component of the wave is its direction of polarization. Some processes, like reflection, tend to produce light that is polarized in a particular direction. Birefringent materials bend light differently depending on its direction of polarization, while polarizing materials like Polaroid only allow light through that is polarized in one direction.

red shift When a source of light moves towards or away from the observer, this has an effect on the wavelength of the light (or its energy, if considering photons). Moving towards the observer increases the energy, shifting the colour of the light up the electromagnetic spectrum to a shorter wavelength (known as a blue shift), while moving away reduces the energy, shifting the colour down the electromagnetic spectrum to a longer wavelength, known as a red shift.

Schrödinger's equation The quantum pioneer Erwin Schrödinger produced an equation that describes the progress of a quantum system over time. Rather than providing an absolute value, like equations derived from Newton's laws, Schrödinger's wave equation (or, to be precise, the square of its result), plots out over time the probability of finding a quantum particle at any location.

visual cortex A part of the cerebral cortex in the brain that processes visual information from the optic nerves.

vacuum Space containing no matter. An approximation to a vacuum can be created by pumping the air out of a vessel, or in the depth of space.

wavelength The distance in its direction of travel that a wave takes to return to a point in its cycle. Wavelength is inversely related to frequency. In light, the shorter the wavelength, the greater the energy of the photon.

THE ELECTROMAGNETIC SPECTRUM

the 30-second theory

Light can be considered as a wave of interacting electricity and magnetism that travels through space, but in fact it is just part of what we call the electromagnetic spectrum. In terms of wavelength, the electromagnetic spectrum goes from the longest radio waves to the shortest gamma rays, with light being a small part of the entire range. If we were to represent the entire electromagnetic spectrum by a piano keyboard, the part corresponding to light would be less than a single key on the keyboard. It was James Clerk Maxwell who showed in the mid-1800s that light was just one form of electromagnetic radiation, the part to which our eyes are sensitive. In 1800 the astronomer William Herschel accidentally discovered what we now call the infrared, and the following year the ultraviolet was accidentally discovered by Johann Wilhelm Ritter. X-rays and gamma rays were discovered in the 1890s. The shorter the wavelength, the more energetic the radiation: so gamma rays, which have the shortest wavelength, have the most energy and are very dangerous. All electromagnetic radiation travels at the speed of light, so radio waves, for example, travel at this speed too.

RELATED TOPICS
See also
SPEED OF LIGHT
page 52

ELECTROMAGNETISM
page 80

3-SECOND BIOGRAPHIES
WILLIAM HERSCHEL
1738–1822
German-born musician and astronomer who discovered the infrared in 1800

JAMES CLERK MAXWELL
1831–79
Scottish theoretical physicist who showed that light was an electromagnetic wave

WILHELM RÖNTGEN
1845–1923
German physicist who discovered X-rays in 1895

30-SECOND TEXT
Rhodri Evans

Maxwell (top), Herschel (centre) and Röntgen (bottom) made key breakthroughs in our understanding of the electromagnetic spectrum.

3-SECOND THRASH
The electromagnetic spectrum includes a whole range of electromagnetic waves, from radio to gamma rays, of which visible light is a tiny part.

3-MINUTE THOUGHT
Electromagnetic waves are produced by a varying electric field and magnetic field at right angles to one other. A varying electric field produces a magnetic field, and a varying magnetic field produces an electric field. Thus, electromagnetic waves self-propagate through space and can travel from one edge of the universe to the other. The cosmic microwave background radiation, for instance, has been travelling for over 13 billion years.

Gamma rays | X-rays | Ultra violet | Infrared | Radio waves Radar TV FM

400 nm

700 nm

COLOUR

the 30-second theory

There are more than a dozen causes of colour, and that's even before you start to think about how the brain processes the light that reaches the eye. As a sensation, colour is as much a matter of psychology and physiology as it is of physics. But it starts with light. When our eyes receive light of different intensities at the wavelengths spanning the visible spectrum (from around 400 to 700 nanometres), the brain generally interprets the light signal as being coloured. The question is then what processes can reduce the intensities of some of the wavelengths in white sunlight to produce this sensation. A common one is absorption. Substances can absorb some wavelengths more than others, ultimately because the photons have just the right energy to boost electrons from one quantum energy state to another. Because chlorophyll molecules absorb red and blue light, the reflected light makes grass appear green. Another cause of colour is light scattering. How strongly small particles and molecules scatter light depends on their size and the wavelength of the light. Molecules in air scatter blue light most strongly, so it seems to come from all directions: the sky looks blue. Interference of reflected light waves, meanwhile, creates the iridescent blues and greens of butterfly wings and insect cuticle.

RELATED TOPICS
See also
ELECTROMAGNETIC
SPECTRUM
page 36

PHOTONS
page 40

3-SECOND BIOGRAPHIES
JOHANN WOLFGANG VON
GOETHE
1749–1832
German writer and naturalist
who opposed Newton's theory
of light and colour

THOMAS YOUNG
1773–1829
English scientist who explained
wave interference and the
basis of colour vision

MICHEL-EUGÈNE CHEVREUL
1786–1889
French chemist whose colour
theory and ideas on colour
contrast influenced artists

30-SECOND TEXT
Philip Ball

Our perception of colour depends on the response of eye and brain to differing intensities of light.

3-SECOND THRASH
We perceive colour when visible light reaching our eyes is more intense at some wavelengths than others.

3-MINUTE THOUGHT
Because the spectrum of light reflected from objects to our eyes varies under different illumination – at midday and dusk, say – our visual system has evolved a means of correcting for these variations to preserve the perceived colour, so that a red apple still looks red, for example. This phenomenon, called colour constancy, involves specialized neurons in the primary visual cortex that recalibrate the signal from wavelength-sensitive retinal cone cells according to its context.

PHOTONS

the 30-second theory

The photon is a 'packet' of electromagnetic radiation. In quantum theory the electromagnetic field consists of photons, and the electromagnetic force arises when two particles exchange one or more photons. Until the end of the 19th century, light was thought to be a wave. Then in 1900 German physicist Max Planck introduced the notion that electromagnetic radiation is not a continuous stream, but occurs in individual packets or quanta known as photons. The energy of these photons is proportional to the frequency of the electromagnetic radiation, making the energies of photons with the highest frequency the greatest. The constant of proportionality h, which links the energy of photons E to their frequency ν in the simple formula $E=h\nu$ is known as Planck's constant. Albert Einstein showed that Planck's hypothesis that light is made up of photons explains a puzzling feature of the photoelectric effect, namely that when light hits a metal, the brightness of the light determines the numbers of electrons emitted, but not their energy. This is explained if light consists of photons, because the brighter the illumination is, the more photons there are to act as projectiles to kick electrons out of the metal.

RELATED TOPICS
See also
QUANTA
page 58

WAVE/PARTICLE DUALITY
page 60

SCHRÖDINGER'S EQUATION
page 62

QED
page 68

3-SECOND BIOGRAPHIES
MAX PLANCK
1858–1947
German physicist who proposed that light should be considered as quanta

ALBERT EINSTEIN
1879–1955
German-born physicist whose 1921 Nobel Prize for physics was for identifying the role of the photon in the photoelectric effect

30-SECOND TEXT
Frank Close

Planck's insight that electromagnetic radiation might take the form of quanta rather than a wave inspired Einstein.

3-SECOND THRASH
In the quantum theory of light, electromagnetic waves appear as a staccato burst of massless particles called photons.

3-MINUTE THOUGHT
The idea of photons seems to contradict the behaviour of light as a wave – such as diffraction or interference, where two light beams created by a double slit can cancel one another. This classic experiment was used to demonstrate that light is a wave, but in modern experiments individual photons can be sent through and an interference pattern still builds up. An explanation would be provided by Schrödinger's wave equation.

REFLECTION

the 30-second theory

Light striking a surface can be absorbed, reflected or (if the material is somewhat transparent) transmitted. In simple terms, reflection is light rebounding, almost like a squash ball hitting the wall. If some wavelengths of white light falling on the surface of an object are absorbed, then the component of visible light that is reflected makes the object look coloured. The angle between an incident light ray and the direction perpendicular to the surface is equal to the angle at which it is reflected: shine a ray at 45 degrees to the surface and it is reflected at that angle, too. If the surface is very smooth, like a mirror or still water, then the reflected light creates a perfect (but reversed) image of the incident light. This is called specular reflection. But if the surface is rough, like frosted glass, the rays bounce off in all directions and the image is lost; this is known as diffuse reflection. Although reflection is well described by classical optics, a full explanation involves the quantum theory of interactions between light and matter, called quantum electrodynamics. Here reflection is understood as re-radiation of light from excited atoms at the surface, and the angle of reflection is that at which the radiated waves all reinforce each other by interference.

3-SECOND THRASH
Reflection comes about when light bounces off surfaces, and may be either specular (mirror-like) for smooth surfaces or diffuse for rough ones.

3-MINUTE THOUGHT
Why are left and right reversed in a mirror, but up and down are not? This seems like it ought to be a simple question to answer. But it has provoked furious debate, even in recent times. The usual answer, given by physicist Richard Feynman, is that it's not left and right that are reversed, but front and back: your nose, previously pointing north (say), points south in the mirror.

RELATED TOPICS
See also
COLOUR
page 38

REFRACTION
page 44

3-SECOND BIOGRAPHIES
ROGER BACON
c.1214–1292
English philosopher who referred to 'laws of reflection and refraction'

AUGUSTIN-JEAN FRESNEL
1788–1827
French physicist who first wrote equations on how light is reflected and refracted

RICHARD FEYNMAN
1918–88
American physicist who won the Nobel Prize for QED, the quantum theory of the interaction of light and matter

30-SECOND TEXT
Philip Ball

In modern cities we get plenty of chances to ponder light reflection off mirror-like surfaces.

REFRACTION

the 30-second theory

3-SECOND THRASH
Refraction is the bending of light as it passes from one medium to another with a different refractive index – for example, from air to water or glass.

3-MINUTE THOUGHT
Some substances, such as the mineral calcite, have different refractive indices in different directions. Such materials are said to be birefringent. Light rays passing through them can follow different paths if they have different polarization (different orientations of their oscillating electromagnetic fields), leading to the formation of double images: double refraction. Birefringence is exploited in liquid-crystal displays, in which molecules aligned in particular directions can appear lighter or darker depending on how they influence polarized light.

Contrary to common belief, the speed of light isn't constant. Light travels more slowly through glass or water than through a vacuum (or air), and this makes a ray change direction when it crosses from one to the other. The effect is called refraction, and it is what causes the distortion in the appearance of objects when they are immersed in water. The ratio of the speed of light in a vacuum to that in another medium is called the medium's refractive index: for all normal substances it is greater than 1. Water has a refractive index of 1.33, and glass, about 1.5. The greater the refractive index, the more light is bent as it enters (or exits). The reason for the bending is that light follows the quickest path between two points: by bending as it enters a slower medium, it can follow a quicker route to a given point than if it got there in a straight line. The angle of refraction of light depends on its wavelength, a phenomenon called dispersion. This is what creates rainbows, as light of different colours is separated by refraction in (and reflection from) raindrops. Refraction also lies behind other 'tricks of the light' such as mirages, which are caused by the different refractive indices of cool and warm air.

RELATED TOPICS
See also
REFLECTION
page 42

POLARIZATION
page 48

PRINCIPLE OF LEAST ACTION/
TIME
page 50

SPEED OF LIGHT
page 52

3-SECOND BIOGRAPHIES
WILLEBRORD SNELLIUS
(SNELL)
1580–1626
Dutch astronomer who deduced the relationship between angle of refraction and relative velocities of light in the transmitting media

THOMAS YOUNG
1773–1829
English scientist who coined the term 'refractive index'

30-SECOND TEXT
Philip Ball

Light rays bending in water cause rainbows and distorted views of underwater objects.

22 September 1791
Born in London, near present-day Elephant and Castle

1805
Apprenticed to the bookbinder George Riebau

1813
Becomes laboratory assistant at the Royal Institution

1821
Undertakes electromagnetic rotation experiment (homopolar motor)

1824
Elected a fellow of the Royal Society

1825
Appointed director of the laboratory at the Royal Institution

1826
Inaugurates the Royal Institution Christmas lectures for children

1831
Discovers electromagnetic induction and constructs an electrical generator

1832
Formulates his two laws of electrolysis

1833
Becomes Fullerian professor of chemistry at the Royal Institution

1845
Discovers a relationship between magnetism and light

1848
Granted use of a grace and favour house at Hampton Court

25 August 1867
Dies, at Hampton Court, near London

MICHAEL FARADAY

The son of a blacksmith, Michael Faraday was born in what is now south London in 1791. He received a basic schooling, and by the age of 14 was apprenticed to a bookbinder. Faraday took every opportunity to devour the contents of the books – especially on a subject that fascinated him, such as electricity or chemistry. Later, he continued his self-education by attending public lectures on scientific subjects, including those by Humphrey Davy at the Royal Institution. In 1812, as he was coming to the end of his apprenticeship, Faraday sent a bound copy of his lecture notes to Davy himself, in the hope of securing a job at the Royal Institution. There was no vacancy at the time, but when Davy fired an assistant a few months later he remembered the young man's application. Faraday was appointed to the post of chemical assistant at the Royal Institution on 1 March 1813.

Faraday proved an outstanding experimentalist, and before long his achievements eclipsed those of Davy himself. In 1825, after Davy's retirement, Faraday took over as director of the laboratory and then in 1833 he became the Royal Institution's first professor of chemistry. The title was a misnomer, however, since Faraday's most important achievements were in physics. His first great discovery had come in 1821, with a phenomenon he described as 'electromagnetic rotation' – effectively the world's first electric motor. The peak of his creativity came in the years 1831 and 1832, with his work on electromagnetic induction, the construction of a simple electrical generator and the formulation of his laws of electrolysis. Faraday's genius lay in his ability to bring together seemingly disparate branches of science: electricity and magnetism, electromagnetism and motion, chemistry and electricity.

As well as being a great scientist, Faraday was a consummate popularizer. He continued Davy's successful public lectures, and gained a reputation as one of London's most entertaining speakers; his admirers included Charles Dickens and members of the Royal Family. In later years Faraday acted as a scientific advisor to the government on issues ranging from lighthouses to mining accidents. During the Crimean War he was asked to look into the use of poison gas as a weapon, but he refused to do so on ethical grounds. His sense of humility led him to turn down numerous honours, including a knighthood. He died in 1867, a few weeks short of his 76th birthday.

POLARIZATION

the 30-second theory

If you think of light as the interplay between an electrical and a magnetic wave at right angles to each other, the electrical wave ripples side to side in a particular direction – this direction is the light's polarization. (It's an arbitrary decision to use the direction of the electric wave.) For those who prefer to think of photons, each photon has a direction associated with it at right angles to its direction of travel, which is its polarization. An ordinary light source like the Sun emits photons with all directions of polarization, but some materials act as filters, only allowing light with a particular direction of polarization through. This was first observed when light was passed through a crystal called Iceland spar, which has different refractive indices for two directions of polarization, so produces two images of something seen through it. Reflected light tends to have more photons polarized in one direction, making it possible for polarizing sunglasses to cut out glare. LCD screens use two polarizing filters at right angles, either side of the liquid crystal. The filters stop light passing through, but when a current is passed across the crystal it rotates the direction of polarization, allowing light to penetrate.

RELATED TOPICS
See also
THE ELECTROMAGNETIC
SPECTRUM
page 36

PHOTONS
page 40

3-SECOND BIOGRAPHIES
ERASMUS BARTHOLIN
1625–98
Danish experimenter who made the first scientific investigation of Iceland spar

AUGUSTIN-JEAN FRESNEL
1788–1827
French engineer and scientist who linked polarization with the direction in which a light wave oscillates

EDWIN LAND
1909–91
American engineer who invented the polarizing material Polaroid

30-SECOND TEXT
Brian Clegg

Polarization is put to everyday use in polarizing sunglasses and LCD screens – and may have a future role in fibre optics.

3-SECOND THRASH
Each photon (or wave) of light has a direction at right angles to its motion associated with its changing electric field, defining its polarization.

3-MINUTE THOUGHT
Conventional 'linear' polarization is in a fixed direction, but light can also have 'circular polarization', where the direction of polarization rotates as the light passes along. By modulating the polarization, rather than modulating the intensity of light, as currently used in fibre optics, it may be possible to double the amount of information transmitted. (Note this is a different effect from experiments that produce light with a rotating phase, a separate property from polarization.)

PRINCIPLE OF LEAST ACTION/TIME

the 30-second theory

3-SECOND THRASH
Light travels along the route that minimizes the time needed to cover a distance, which means that when passing from air into glass, where it's slower, the light bends inwards.

3-MINUTE THOUGHT
The principle of least action fascinated Richard Feynman, presenting, as it does, a totally different way of looking at nature. It was the starting point for his PhD thesis which proposed that by drawing every possible 'world line' describing how a particle can get from A to B and attaching probabilities to these different lines it was possible to provide a much more intuitive description of a quantum particle's behaviour than earlier mathematical formulations.

The principle of least action shows nature to be lazy. For instance, the trajectory of a ball through the air takes the route that minimizes the difference between the ball's kinetic and potential energies. In the 17th century, Pierre de Fermat applied a variant of the principle, the principle of least time, to explain refraction – the bending of light when it passes, say, from air to glass. The principle says that the light will take the quickest route from A to B. In a single medium this is a straight line. But light travels more slowly in glass than it does in air. Because of this, it is quicker to spend longer in air to achieve a reduction in the time spent in glass. So a ray that travels further in air, then bends in towards the perpendicular to travel less far in glass is quicker than a straight line. This is sometimes called the 'Baywatch principle' because the same concept applies to a lifeguard on a beach. The quickest way to reach a drowning person in the water is not to head straight towards them, but to head off at angle so that more time is spent running on the beach and less time in (slower) swimming through water.

RELATED TOPICS
See also
REFRACTION
page 44

MOVEMENT, SPEED & VELOCITY
page 98

KINETIC ENERGY
page 122

POTENTIAL ENERGY
page 124

3-SECOND BIOGRAPHIES
PIERRE DE FERMAT
1601–65
French mathematician who first applied the principle of least action to light

RICHARD FEYNMAN
1918–88
American physicist who extended the principle of least action to quantum physics

30-SECOND TEXT
Brian Clegg

The quickest route may not be the shortest. A lifeguard moves quicker across the beach than through water.

SPEED OF LIGHT

the 30-second theory

Light travels very, very fast, but not quite instantaneously. Galileo tried to measure the speed of light using lanterns on hills a few kilometres apart, but failed – and he concluded that it got from A to B instantly. The first indication that light has a finite speed came in the late 1600s, when Danish astronomer Rømer realized that the timings of the orbits of Jupiter's moons were different when Earth was closer to Jupiter than when it was further away. In the mid-1800s experiments by Frenchmen Fizeau and Foucault produced values for the speed of light that came within a few per cent of the currently accepted 300,000 km/sec (or 186,000 miles per second). In 1865 Maxwell showed that light was a form of electromagnetic radiation, as its speed was identical to that of an electromagnetic wave predicted by known data on electricity and magnetism. In 1905 Einstein suggested that the speed of light was a fundamental constant of nature and all observers, no matter how quickly they were travelling, would measure the same speed for it. He also argued that nothing could travel faster than light, and that as we approach the speed of light lengths get shortened, our time as seen by an external observer runs slower and masses increase.

3-SECOND THRASH
Light travels very quickly, at 300,000 km/sec, and Einstein argued that nothing can travel faster than light.

3-MINUTE THOUGHT
Because of the finite speed of light, when we look into the night-time sky we are seeing back in time. Light from Sirius, the brightest star in the night-time sky, takes 8.5 years to reach us, light from the North Star takes about 400 years. Light from the most distant galaxies we can see has taken around 13 billion years – we see them as they were when the universe was very young.

RELATED TOPICS
See also
PHOTONS
page 40

ELECTROMAGNETISM
page 80

SPECIAL RELATIVITY
page 112

3-SECOND BIOGRAPHIES
OLE RØMER
1644–1710
Danish astronomer; the first person to get a measurement for the speed of light

LÉON FOUCAULT
1819–68
French physicist best known for the Foucault pendulum

JAMES CLERK MAXWELL
1831–79
Scottish theoretical physicist who showed that light was an electromagnetic wave

30-SECOND TEXT
Rhodri Evans

Rømer saw that orbital times for Jupiter's moons differ depending on Earth's distance from the giant planet.

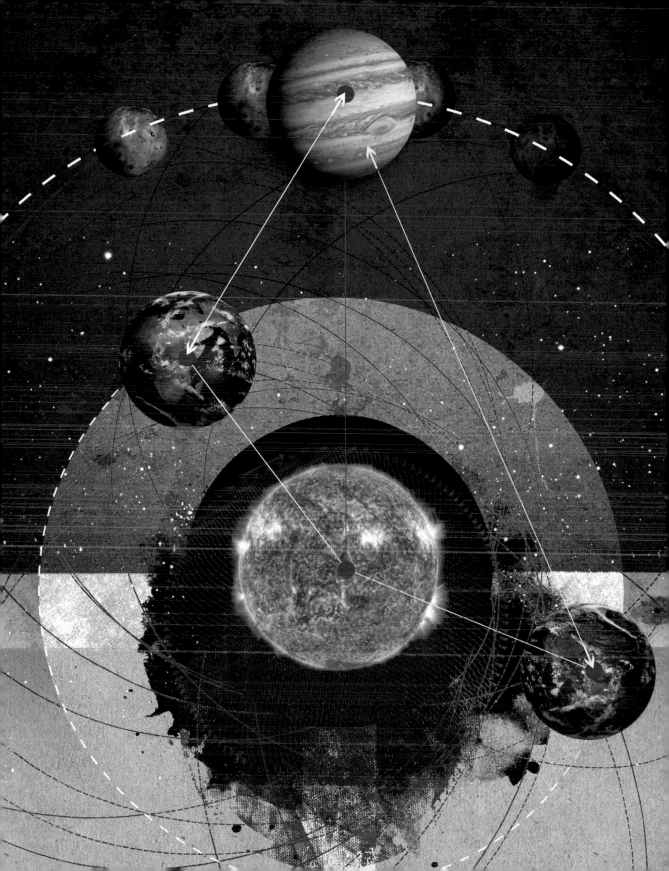

QUANTUM THEORY

antiparticles Usually particles of antimatter – similar to matter but typically with the opposite charge. Each matter particle has an equivalent antiparticle. So, for instance, the electron's antiparticle is the positron. When matter and antimatter come together the particles can annihilate to produce the equivalent amount of energy in the form of photons. Under some definitions of antiparticle, a photon is its own antiparticle.

blackbody A physical object that absorbs all electromagnetic radiation that reaches it. At a constant temperature, a blackbody emits a spectrum of light that depends solely on its temperature.

Dirac equation An equivalent of the Schrödinger equation that takes into account the effects of special relativity, developed by British physicist Paul Dirac. The equation works for particular types of particle, like the electron. The equation implied the existence of the positron long before it was detected.

electrolysis Using an electric current to drive a chemical reaction. The electric current adds electrons to what would otherwise be positively charged ions (atoms missing electrons) and removes electrons from negatively charged ions. One of the best-known examples is the electrolysis of water, producing hydrogen gas and oxygen gas.

Heisenberg's uncertainty principle A result from quantum physics that means that the better you know a quantum particle's position, the less accurately you can know its momentum, and vice versa. The same goes for a system's energy and time.

matrices The plural of 'matrix', matrices are a set of numbers or mathematical expressions laid out in rectangular form. There are special rules for the multiplication of matrices that means that AB does not necessarily equal BA.

momentum In classical physics, the mass of an object times its velocity. In quantum physics, momentum is Planck's constant divided by the wavelength associated with the quantum particle, which applies to both massive particles and those without mass, like a photon.

photon A massless quantum particle of light. Light can be described as a wave, a particle or a disturbance in an electromagnetic field. All of these are models that help us understand it – light itself is just light. Describing light as a particle is helpful when dealing with the interaction between light and matter (QED or quantum electrodynamics), and was first established when Einstein described the way that energetic photons can knock electrons out of metals, producing an electric current in the photoelectric effect. The energy of a photon is equivalent to the light's colour – its frequency or wavelength when thought of as a wave. The photon is the carrier particle of the electromagnetic force: when two objects interact electrically or magnetically, photons travelling between the objects carry the force.

quanta Plural of quantum. 'Quanta' was first used to describe packets or particles of light when it was discovered that light sometimes behaved as a collection of discrete objects. 'Quanta' now refers to all objects small enough to be subject to quantum physics.

quantum state A quantum system, which can be for one or more quantum particles, has a collection of numbers that defines the state that it is in. Typically a quantum property like spin does not have a single definitive value before it is measured, but rather is in a quantum state, which may be, for instance, 40 per cent up and 60 per cent down.

[quantum] spin The spin of a quantum particle is one of its properties. Although it was modelled on angular momentum, it is not really about rotation. It comes in values that are multiples of ½ and has a direction that is quantized. So, for instance, if you measure the spin of a particle it will always be either 'up' or 'down' in the direction measured.

Schrödinger's equation The quantum pioneer Erwin Schrödinger produced an equation that describes the progress of a quantum particle over time. Rather than providing an absolute value, like equations derived from Newton's laws, Schrödinger's wave equation (or, to be precise, its square), plots out over time the probability of finding a quantum particle at a particular location.

QUANTA

the 30-second theory

Quanta is the plural of

'quantum', which in physics is the minimum amount of any physical entity that is involved in an interaction. The idea that nature is quantized was first suggested in 1900 by Max Planck. He proposed that the spectrum of blackbodies could be explained if it was assumed it was only possible to emit light with certain energies, in bundles he called 'quanta' – the energy of each quantum being dependent on the light's frequency. Five years later Albert Einstein was able to explain the photoelectric effect, whereby electrons are released from the surface of certain metals when light shines on them, by arguing that the incoming light was absorbed by the electrons in discrete energy bundles, the same quanta that Planck had proposed for the emission of light. After the discovery of the atomic nucleus in 1913, Niels Bohr suggested that the orbits of electrons about the nucleus were quantized and could only take certain values. These quantized orbits allowed Bohr to explain the spectrum of hydrogen gas, with photons of specific frequencies being emitted when electrons jumped from higher to lower orbits. In the late 1920s Erwin Schrödinger and Werner Heisenberg independently developed what we now call 'quantum theory', which explained Bohr's idea of quantized orbits for electrons.

RELATED TOPICS
See also
PHOTONS
page 40

WAVE/PARTICLE DUALITY
page 60

THE UNCERTAINTY PRINCIPLE
page 64

3-SECOND BIOGRAPHIES
MAX PLANCK
1858–1947
German physicist who was the first to suggest that light was quantized

ERWIN SCHRÖDINGER
1887–1961
German physicist who developed the wave mechanics formulation of quantum theory

WERNER HEISENBERG
1901–76
German physicist who developed the matrix mechanics formulation of quantum theory

30-SECOND TEXT
Rhodri Evans

Bohr's paper On the Constitution of Atoms and Molecules *outlined his new theory in 1913.*

3-SECOND THRASH
The energy of radiation and of subatomic particles like electrons and protons can only come in discrete packets, which we call quanta.

3-MINUTE THOUGHT
The idea that energy comes in discrete packets – quanta – was one of the great revolutions of 20th-century physics. The revolution was started by Max Planck in 1900 and culminated in the late 1920s with quantum mechanics, which tells us that not only is energy quantized but also there is an inherent uncertainty to all measurements.

n1

n2

n3

n4

n5

WAVE/PARTICLE DUALITY

the 30-second theory

RELATED TOPICS
See also
QUANTA
page 58

SCHRÖDINGER'S EQUATION
page 62

THE UNCERTAINTY PRINCIPLE
page 64

3-SECOND BIOGRAPHIES
MAX PLANCK
1858–1947
German physicist who proposed that light and heat energy was parcelled into tiny packets – quanta

CLINTON DAVISSON
1881–1958

LESTER GERMER
1896–1971

& GEORGE PAGET THOMSON
1892–1975
Two American physicists and one English scientist who showed that electrons exhibit diffraction, a property of waves

3-SECOND THRASH
Quantum mechanics describes a strange world in which apparent waves, including light waves, behave as particles and apparent particles, such as electrons, behave as waves.

3-MINUTE THOUGHT
Wave/particle duality extends beyond individual particles. Atoms and even larger molecules have been shown to behave like waves. Physicists believe that every object has a wavelength associated with it – even you – and that increasingly massive objects have ever-smaller wavelengths. Fortunately, we do not see the effects of this in everyday life because rocks, cars, people and planets are just so massive compared with individual particles that their wavelengths are imperceptibly tiny.

We think of light as being a wave and of the building blocks of atoms – electrons, neutrons and protons – as being particles. But in the confusing world of quantum mechanics, waves such as light can behave like particles and particles such as electrons can behave like waves. This phenomenon is called wave/particle duality and it has been demonstrated in many experiments. The behaviour of light in certain circumstances can only be explained if light is understood as being packaged into discrete parcels or quanta – that is, particles. A quantum of light is called a photon: it has momentum just like a particle and a wavelength and a frequency associated with waves. Similarly, some behaviour seen in electrons can only be explained if they are thought of as waves. Electrons and all other particles have a wavelength and a frequency, as well as the momentum expected of a particle. Modern digital cameras have components that are easier to explain by treating light both as a wave (when it is focused by the lens) and a particle (when a photon of light hits the detector chip and releases an electron). Similarly, both the wave and particle properties of electrons are exploited in electron microscopes.

30-SECOND TEXT
Leon Clifford

According to physicists, even Einstein and Max Planck had wavelengths associated with them.

SCHRÖDINGER'S EQUATION

the 30-second theory

The equations of classical mechanics, which describe how a system changes over time, break down in the quantum world. A different mathematical approach is needed to describe how a quantum system evolves. German physicist Erwin Schrödinger provided the solution to this problem in 1926 with what is now called the Schrödinger equation. This equation actually describes the changes in what is known as the wave function of a quantum system. The wave function captures all the information needed to describe a quantum system fully. The Schrödinger equation describes how the probability of locating a particle or a system of particles evolves in a wavelike manner and so provides some insight into wave/particle duality – the property of particles to behave like waves and waves to behave like particles. Mathematically, the Schrödinger equation is based on algebra and calculus. Quantum mechanics can also be described using the mathematics of matrices. Both approaches are equivalent. The Schrödinger equation was a significant advance on the mathematics of classical physics and provided a foundation that was built on by English physicist Paul Dirac and helped to create quantum electrodynamics (QED), a key plank in modern physics.

RELATED TOPICS
See also
QUANTA
page 58

WAVE/PARTICLE DUALITY
page 60

THE UNCERTAINTY PRINCIPLE
page 64

QED
page 68

3-SECOND BIOGRAPHIES
ERWIN SCHRÖDINGER
1887–1961
Nobel Prize-winning Austrian physicist, creator of the Schrödinger equation

WERNER HEISENBERG
1901–76

& RICHARD FEYNMAN
1918–88
German and American physicists who found different approaches to describing how quantum systems evolve

30-SECOND TEXT
Leon Clifford

Schrödinger tested his equation by applying it to the structure of the hydrogen atom.

3-SECOND THRASH
Erwin Schrödinger did for quantum mechanics what Isaac Newton did for classical mechanics: he developed a simple equation that describes how a quantum system evolves.

3-MINUTE THOUGHT
Just as the equations of classical mechanics break down at the quantum level so the basic equation of quantum mechanics – the Schrödinger equation – breaks down at near-light speeds. Physicist Paul Dirac solved this problem by combining the mathematics of the very fast – the theory of special relativity – with the mathematics of the very small described by Schrödinger to create the Dirac equation, a relativistic version of the Schrödinger equation for some types of particle.

THE UNCERTAINTY PRINCIPLE

the 30-second theory

This fundamental principle of quantum theory – formulated by German theoretical physicist Werner Heisenberg in 1927 – states that you cannot simultaneously measure the position and momentum of a particle with perfect accuracy or exactly specify its energy at a specific instant. The more precise the measurement of one property, the less precisely the other can be measured or controlled. The effect of this phenomenon is so small that it can be ignored in everyday affairs, but it is dramatic for subatomic particles. This uncertainty is an intrinsic property of nature, not simply a failure in the measuring apparatus. One consequence is that the total energy of a particle can fluctuate by some amount E for a short time t as long as the product of E times t does not exceed Planck's constant divided by 4π. This in turn means that energy conservation can be put on hold for very short time spans. Particles in such a state are known as virtual particles. The exchange of virtual photons between particles gives rise to the electromagnetic force, according to QED.

3-SECOND THRASH
If you know where something is, you cannot also know precisely where it's going; and energy can be 'borrowed' – but only for an instant.

3-MINUTE THOUGHT
The uncertainty principle is the reason why particle accelerators, such as the Large Hadron Collider, are so huge. To probe distances thousands of times smaller than a proton requires beams of particles with energies that are trillions of times greater than found at room temperature. To speed particles to such energies requires big accelerators because present technology limits the rate at which particles can be energized.

RELATED TOPICS
See also
PHOTONS
page 40

QED
page 68

THE WEAK NUCLEAR FORCE
page 88

THE STRONG NUCLEAR FORCE
page 90

3-SECOND BIOGRAPHIES
NIELS BOHR
1885–1962
Danish physicist who worked closely with Heisenberg

WERNER HEISENBERG
1901–76
German physicist who was the father of the uncertainty principle

30-SECOND TEXT
Frank Close

The uncertainty principle – you cannot accurately measure a particle's position and momentum at the same time – is one of physics' most celebrated ideas.

TUNNELLING

the 30-second theory

The phenomenon of quantum tunnelling explains how subatomic processes such as nuclear fusion and radioactivity can overcome what appear to be energy barriers that should be too large for them to happen. For example, in radioactive beta decay a neutron changes into a proton, an anti-neutrino and an electron, with the anti-neutrino and the electron being spat out of the nucleus at high speed. Using classical physics, the electron should not be able to escape the nucleus as the electromagnetic attraction between it and the positively charged nucleus is too strong. To explain how it can escape, the idea of tunnelling was introduced; the electron is able to surmount an energy barrier by 'tunnelling' through it, rather like a train getting to the other side of a mountain by tunnelling through rather than going over it. The tunnelling is achieved by virtue of Heisenberg's uncertainty principle: in the example above, the electron is able to borrow some energy for a brief period of time and this borrowed energy is sufficient to get it over the energy barrier that would otherwise block its escape. Although it only operates over tiny distances of 1–3 nanometres (nm) and less, tunnelling can have macroscopic effects, such as being the major source of power drain in mobile phone electronics.

RELATED TOPICS
See also
WAVE/PARTICLE DUALITY
page 60

THE UNCERTAINTY PRINCIPLE
page 64

THE WEAK NUCLEAR FORCE
page 88

3-SECOND BIOGRAPHIES
MAX BORN
1882–1970
German physicist who realized that tunnelling had a wider use beyond nuclear processes

GEORGE GAMOW
1904–68
Russian-born physicist who used tunnelling to explain radioactive alpha decay

BRIAN JOSEPHSON
1940–
Welsh physicist who did pioneering work on the role of tunnelling in superconductors

30-SECOND TEXT
Rhodri Evans

'Borrowed' energy fuels tunnelling at extremely small subatomic distances.

3-SECOND THRASH
Tunnelling allows subatomic particles to borrow energy for a brief period of time and get over energy barriers that would otherwise be too large.

3-MINUTE THOUGHT
The scanning tunnelling microscope (STM), invented in 1981, scans the surfaces of materials and allows individual atoms to be seen. It works by scanning an electrically conducting tip very close to the surface of the material, and by applying a voltage electrons can tunnel in the gap between the tip and the surface. An STM can resolve down to about 0.1 nm in the horizontal x–y direction, and to about 0.01 nm in the vertical z direction.

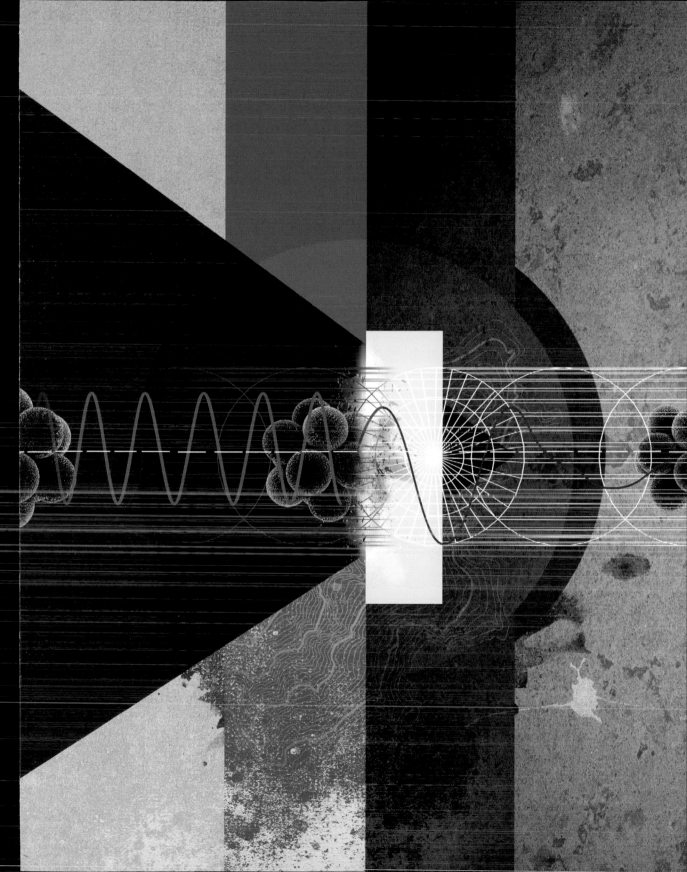

QED

the 30-second theory

Quantum electrodynamics, or QED, is a theory of the electromagnetic force that combines Maxwell's classical theory of electromagnetism, Einstein's special theory of relativity and quantum theory. Maxwell's classical theory of electric currents and electromagnetic waves such as light and radio waves was developed before the discovery of the electron and photon. In 1928 Paul Dirac built a theory of the electron and its interactions with photons that is consistent with special relativity. However, his equation also predicted the existence of antimatter – positrons – and the possibility that an electron and a positron could mutually annihilate in a burst of energy, producing photons, or conversely that photons could convert into a particle and antiparticle such as an electron and a positron. To account for these complexities Dirac developed quantum electrodynamics, which describes the interaction of photons and electric charges, including the effects of matter and antimatter in the electromagnetic field. According to QED, the electromagnetic force between two particles arises as the result of their exchange of one or more photons. The theory is so successful that it describes the magnetic properties of particles such as the electron to an accuracy of about one part in a billion.

RELATED TOPICS
See also
ANTIMATTER
page 28

PHOTONS
page 40

3-SECOND BIOGRAPHIES
PAUL DIRAC
1902–84
English physicist who developed the underlying theory of QED

SIN-ITIRO TOMONAGA
1906–79
JULIAN SCHWINGER
1918–94
& RICHARD FEYNMAN
1918–88
One Japanese and two American physicists who shared the Nobel Prize for work on QED

30-SECOND TEXT
Frank Close

Feynman diagrams represent photons as wobbly lines and electrons or positrons as straight lines, travelling through spacetime.

3-SECOND THRASH
Thanks to the pioneering work of Paul Dirac, QED explains the electromagnetic field in a way that is consistent with both quantum theory and special relativity.

3-MINUTE THOUGHT
QED is one of the most stringently tested theories in physics. Feynman diagrams enable complicated quantum mathematics in QED to be visualized in terms of particles, antiparticles and photons. QED is the inspiration for theories of the strong and weak forces: quantum chromodynamics (QCD) and quantum flavourdynamics (QFD), respectively. Mathematical similarities among these theories have led theorists to search for a grand unified theory of these forces.

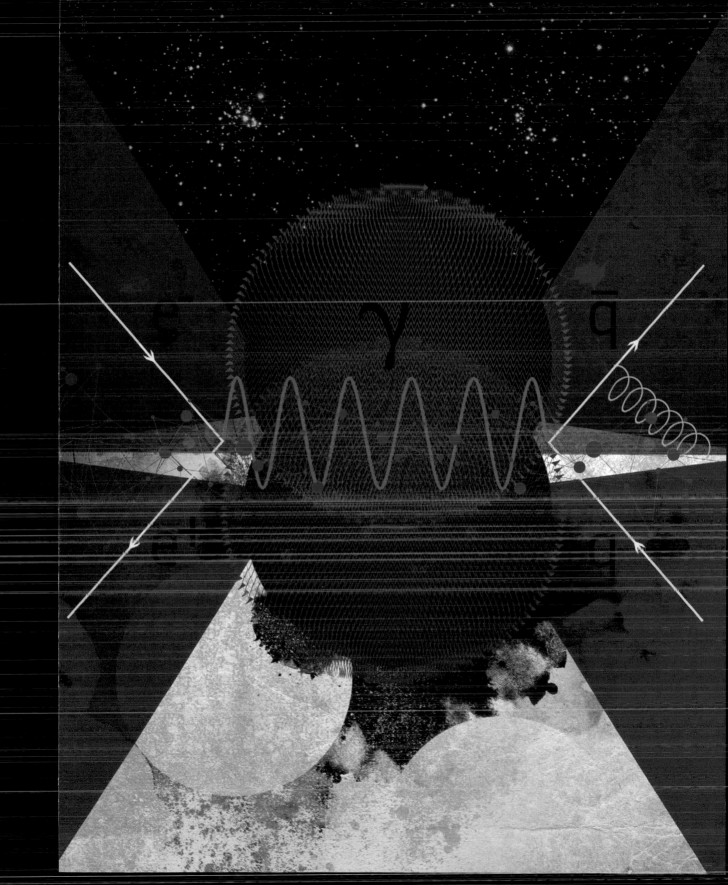

11 May 1918
Born in New York City

1935
Wins a scholarship to MIT, where he majors in physics

1939
Enters Princeton to do his PhD, achieving maximum marks in the entrance exams

1942
Recruited by Robert Oppenheimer to join the Manhattan Project

1945
His childhood sweetheart and wife Arlene dies of tuberculosis

1945
Appointed professor of theoretical physics at Cornell University

1947–49
Does the work on QED that will win him his Nobel Prize for physics

1950
Appointed professor of theoretical physics at Caltech; spends first year on sabbatical in Rio de Janeiro

1960–63
Rewrites and gives the introductory undergraduate physics lectures at Caltech

1965
Awarded the Nobel Prize for physics for his work on QED

1985
Publishes his best-selling memoirs *'Surely You're Joking Mr. Feynman!': Adventures of a curious character*

1986
Uses a simple demonstration to turn around the Rogers Commission, investigating the Challenger space shuttle disaster

15 February 1988
Dies in Los Angeles

RICHARD FEYNMAN

Richard Feynman grew up in Far Rockaway, a suburb of New York City, the son of a uniform salesman. By his teenage years he was devouring maths books aimed at adults, and won several maths competitions. He won a scholarship to MIT, where he majored in physics, before applying to Princeton for his PhD. Feynman achieved maximum marks in his Princeton maths and physics entrance exams, a feat never before accomplished. At Princeton he worked on quantum mechanics under the supervision of John Archibald Wheeler, and as he was finishing his PhD he was recruited by Robert Oppenheimer to join the Manhattan Project in Los Alamos.

At Los Alamos he worked on neutron calculations, which were crucial in creating a chain reaction, but he primarily oversaw the team of human 'computers' who were crunching the numbers in the complex calculations. This work was taking too much time, so Feynman devised parallel computing techniques that saw his team go from doing three calculations in nine months to nine calculations in three months. At the end of the war, turning down an offer to join Einstein at the Institute for Advanced Study in Princeton, Feynman took up a professorship at Cornell University. In the same year his childhood sweetheart and wife Arlene died of tuberculosis, a loss that left him in a deep depression.

Throughout his five years at Cornell, during which he was one of three physicists developing QED, the theory of the interaction between light and matter particles, other universities tried to attract him, Chicago making him an offer he felt was so generous that he could not possibly accept it. He finally chose Caltech, but spent his first year on sabbatical in Rio de Janeiro – where he took part in the Carnival parade playing percussion. At Caltech, Feynman made major contributions to our understanding of the weak nuclear interaction and on the early theory of what would become known as quarks. In the early 1960s he was asked to rewrite the two-year introductory physics course that all Caltech students had to take – these lectures became so popular that graduate students and fellow faculty-members would attend them, and *The Feynman Lectures on Physics* has gone on to become a classic textbook of introductory physics. He received the Nobel Prize for physics in 1965 for his work on quantum electrodynamics, but he was never comfortable with the award. In later life he would say that he wished he had never been given it. His unparalleled ability to explain complex physics in an entertaining and comprehensible manner led to his becoming a TV celebrity, and his memoirs 'Surely You're Joking Mr. Feynman!', published in 1985, has become one of the best-selling science books ever. Feynman's last major contribution was the public exposure of the O-ring failure in the Challenger space shuttle disaster, demonstrated live on TV on 11 February 1986. He died of cancer in February 1988 in Los Angeles.

Rhodri Evans

ENTANGLEMENT

the 30-second theory

Described by Einstein as 'spooky action at a distance', entanglement is a fascinating phenomenon that occurs in quantum mechanics whereby the quantum states of two or more particles are linked. For example, we can generate two electrons with a total spin of zero (spin is one of an electron's four quantum states), and send them off in different directions so that they have a large separation between them. If one of the electrons is then measured to have a spin of $+\frac{1}{2}$, we know that the other electron must have a spin of $-\frac{1}{2}$. When we measure the quantum state of the first electron the other possible states disappear, and we say that its quantum state has 'collapsed'. This leads to the paradoxical situation in which the second electron somehow 'knows' the quantum state of the first electron instantaneously, even though the distance between them may be large. This apparent paradox, that information has passed between the two electrons instantaneously, is known as the 'Einstein, Podolsky, Rosen paradox' (or the 'EPR paradox').

RELATED TOPICS
See also
SPEED OF LIGHT
page 50

SPECIAL RELATIVITY
page 112

ALBERT EINSTEIN
page 108

3-SECOND BIOGRAPHIES
BORIS PODOLSKY
1896–1966

& NATHAN ROSEN
1909–95
American physicists who together with Einstein were behind the 'EPR paradox'

JOHN BELL
1928–90
Northern Irish physicist who developed Bell's theorem, which made the existence of entanglement's instantaneous connection testable

30-SECOND TEXT
Rhodri Evans

Measuring +½ spin for one particle means we know at once the other must measure –½ spin – no matter how far apart they are.

3-SECOND THRASH
Entanglement is the strange phenomenon in which two particles are able to communicate their quantum state instantaneously to each other, even over vast distances.

3-MINUTE THOUGHT
One of the consequences of quantum entanglement is quantum teleportation. Whilst this isn't quite the teleportation envisaged in TV show *Star Trek*, it provides a way for a unit of quantum information (a 'qubit') to be instantaneously transferred between two separate places. The current record for instantaneous quantum teleportation for photons is 143 km (89 miles), achieved in 2012.

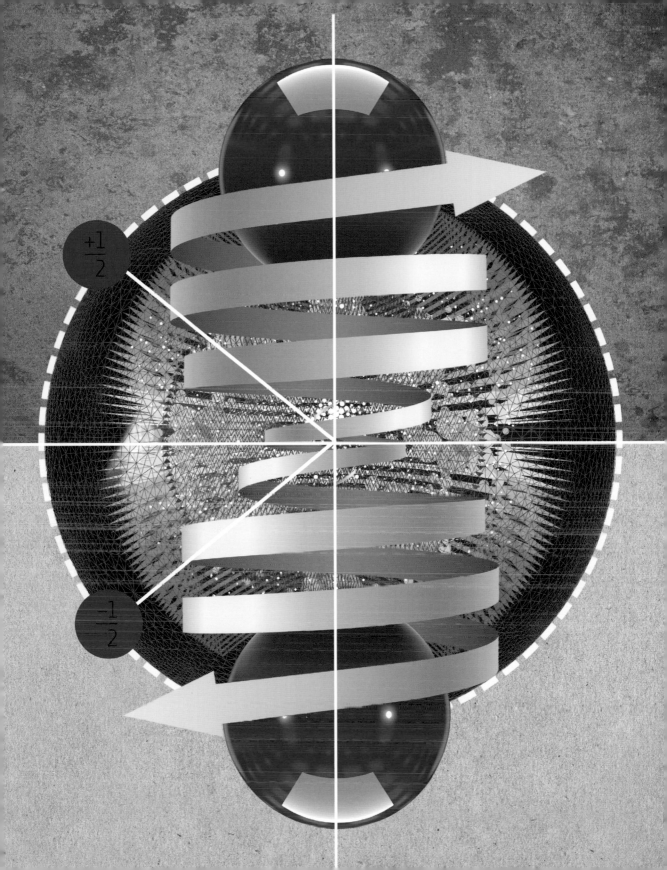

FORCES

electrical disruption In the nucleus of an atom, two forces oppose each other: positively charged protons repel each other electromagnetically, but are attracted to other protons, and neutrons, by the strong force. Because the strong force is very short range, if there are too many protons, the electrical repulsion between the protons can overcome it, disrupting the nucleus and breaking it apart.

Galilean relativity Galileo realized that it wasn't enough to say that something moves, it was also necessary to say what the movement was compared to – hence, relativity. He postulated that in an enclosed boat, moving at a constant speed, it would be impossible to perform an experiment producing different results to the same experiment performed when the boat was not moving. As far as the objects inside the steadily moving boat are concerned, the boat is not moving – it is only an outside observer who sees it moving.

general relativity Einstein first extended Galilean relativity by bringing in the fixed speed of light to produce the special theory of relativity. He then extended relativity to cover acceleration and gravity, producing the general theory of relativity, which treats gravity as a warp in spacetime caused by objects with mass.

neutrinos A neutral or uncharged fundamental quantum particle with an almost undetectable mass that is produced during nuclear reactions. The neutrino was predicted to exist in 1930 to explain the loss of energy during a nuclear reaction, but it was not detected until 1956 because it has very little interaction with matter. Billions of neutrinos from the Sun pass through each of us every second. The name means 'little neutral one'.

Newton's first law of motion Also called the law of inertia. It says that a body will stay at rest or in uniform motion in a straight line (that is, with a fixed velocity) unless a force acts on it.

pions A quantum particle – more properly a 'pi meson'. As a meson, the particle is composed of a quark and an antiquark. Pions are unstable, decaying in a tiny fraction of a second. Pions can be charged or neutral – charged pions usually decay to a muon (an elementary particle like a heavy electron) and a neutrino, while neutral pions decay into photons.

quarks A fundamental quantum particle, with either ⅔ the charge of a proton or ⅓ the charge of an electron. Quarks come in six different 'flavours': up, down, charm, strange, top and bottom. Triplets of quarks make up protons and neutrons, while pairs of a quark and an antimatter quark make up another type of particle, mesons.

radioactive beta decay One of the ways an atomic nucleus can decay is when a neutron is converted to a proton, or a proton to a neutron. (This tends to happen to move towards a more stable nuclear structure.) The result is that the atom becomes a different element, because the element is defined by the number of protons in the nucleus. In beta decay, the nucleus emits either an electron or a positron, as well as a neutrino or antineutrino. When the process was first discovered, the electrons emitted were called beta rays (later beta particles) to distinguish them from the positively charged alpha particles that are also sometimes emitted. The transformation between proton and neutron is due to the weak force.

scalar A property that only has a numerical value, like mass, is a scalar (as opposed to a vector).

spacetime In the special theory of relativity it becomes impossible to treat time and space separately, as each is dependent on the other. Rather than handle them separately, physicists work with the spacetime continuum, treating time as a (special) fourth dimension.

vector A property that has a numerical value *and* a direction is a vector. Velocity, for instance, is made up of speed (a scalar) and the direction of that speed.

FORCE & ACCELERATION

the 30-second theory

3-SECOND THRASH
To accelerate an object we need to apply a force, but how much the acceleration is depends on the body's mass.

3-MINUTE THOUGHT
Gravity was the first force of nature to be understood. Newton's law of universal gravitation went unchallenged until Einstein's general theory of relativity, which since 1915 has replaced it. We now know of three other forces; the electromagnetic force; and the strong and weak nuclear forces, which govern the nuclei of atoms. Unifying these four forces into a single theory is one of the great unsolved problems in physics.

Newton's first law of motion, sometimes called the law of inertia, tells us that to get a body moving or to change its velocity once it is moving (either its speed or its direction of motion) we need to apply a force. His second law of motion tells us how an applied force will change that body's motion, linking the force applied to the mass of the body and its acceleration. In physics, an acceleration can mean either a change in speed or a change in direction (or both). Newton tells us that the acceleration produced will be the force applied divided by the object's mass. This means, for example, that a larger force is required to accelerate a massive body than to accelerate a less massive one – which is why we need a more powerful engine in a large articulated lorry than we do on a motorbike. It also means that a force is required to keep the Earth in orbit about the Sun as, being in orbit, it does not travel in a straight line. Newton realized that this force is gravity – the same force that attracts apples (and us) to the Earth.

RELATED TOPICS
See also
ELECTROMAGNETISM
page 80

THE WEAK NUCLEAR FORCE
page 88

THE STRONG NUCLEAR FORCE
page 90

3-SECOND BIOGRAPHIES
GALILEO GALILEI
1564–1642
Italian natural philosopher who was the first to suggest the concept of inertia

JOHANNES KEPLER
1571–1630
German mathematician, first to realize that planets orbit the Sun in ellipses and not circles

ISAAC NEWTON
1643–1727
English physicist whose *Principia* (1687) provided a mathematical framework for physics for 250 years.

30-SECOND TEXT
Rhodri Evans

Force divided by mass gives acceleration – a change in speed or direction.

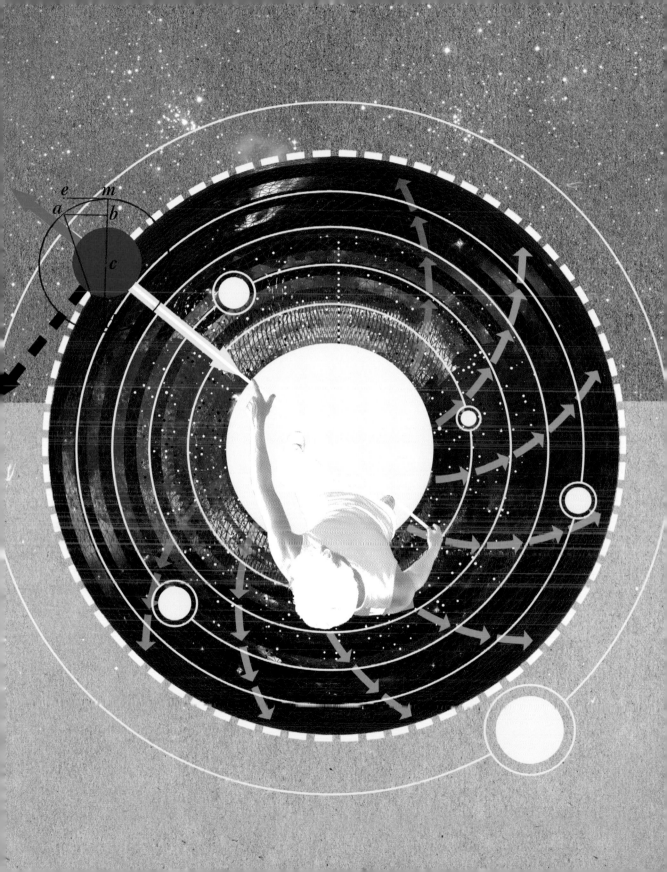

ELECTROMAGNETISM

the 30-second theory

For centuries people had known that magnets and electrically charged objects could attract or repel each other. With the invention of the electric battery in 1800, scientists started investigating the properties of electric currents, which are simply electrical charges moving in wires. It was shown by Ampère that two wires each carrying an electric current could attract or repel each other, depending on whether the currents flowed in the same or in opposite directions. Ørsted showed that a wire carrying an electric current produced a circular magnetic field about it. Faraday investigated these phenomena further, and found that a wire carrying a current in a magnetic field would experience a force upon it, and later he discovered that a wire moving in a magnetic field would have an electric current induced in it. Faraday also showed that a magnetic field could bend light and that a changing current in a wire could induce a current to flow in a nearby wire. In the 1860s Maxwell brought all of these related phenomena together when he derived his laws of electromagnetism, linking electricity and magnetism and explaining the nature of light.

RELATED TOPICS

See also
THE ELECTROMAGNETIC
SPECTRUM
page 36

MICHAEL FARADAY
page 46

JAMES CLERK MAXWELL
page 148

3-SECOND BIOGRAPHIES
ANDRÉ-MARIE AMPÈRE
1775–1836
French physicist

MICHAEL FARADAY
1791–1867
English experimental scientist who clarified and made practical the link between electricity and magnetism

30-SECOND TEXT
Rhodri Evans

3-SECOND THRASH
Electricity, magnetism and light are all part of the same phenomenon known as electromagnetism, which was first fully described by Maxwell in the 1860s.

3-MINUTE THOUGHT
Electromagnetism was the second force of nature to be explained, after gravity. It is the reason that atoms combine to form molecules and why you do not sink through your chair when you sit down. The electromagnetic force is much stronger than gravity: the force between a small magnet and your fridge is strong enough to overcome the gravitational force of the entire Earth pulling down on the magnet.

James Clerk Maxwell (far right) mathematically described electromagnetism with his equations.

GRAVITY

the 30-second theory

RELATED TOPICS
See also
ELECTROMAGNETISM
page 80

THE WEAK NUCLEAR FORCE
page 88

THE STRONG NUCLEAR FORCE
page 90

3-SECOND BIOGRAPHIES
ARISTOTLE
384–322 BCE
Greek philosopher

ALBERT EINSTEIN
1879–1955
German-born physicist, and winner of the 1921 Nobel Prize for physics

KIP THORNE
1940–
American physicist, arguably the foremost living expert in general relativity

30-SECOND TEXT
Rhodri Evans

3-SECOND THRASH
Newton was the first to describe gravity and for most situations his theory suffices, but Einstein replaced Newton's theory in 1915 with a new theory that describes gravity as due to a bending of space and time.

3-MINUTE THOUGHT
Gravity is one of four forces in nature – the others being electromagnetism and the strong and weak nuclear forces. General relativity is currently incompatible with theories that have unified the other three forces, and one of the holy grails of physics is to develop a theory that will unify gravity with them. Possible candidates include string theory, loop quantum gravity and M-theory.

In the 4th century BC, Aristotle argued that objects fell because the elements earth and water sought to be at the centre of the universe. In the late 1600s, nearly 2,000 years later, Newton overthrew this view when he realized the force that causes an apple to fall to the ground is the same as the force that holds the Moon in its orbit. He described how 'universal gravitation' acts between any two bodies. Its strength depends on the bodies' masses multiplied together divided by the square of the distance between them. Newton's law of gravity has been spectacularly successful; using it, scientists were able to predict the position of an unseen planet (Neptune) due to irregularities in the orbit of Uranus, and it allows us to land probes on comets hundreds of millions of kilometres away. In 1915 Einstein published a radical new theory of gravity – general relativity – because he realized that Newton's law of gravity violated his principles of special relativity. General relativity describes gravity in an entirely different way, as a warp in space and time. Mass tells spacetime how much to curve: more massive objects cause more curvature of spacetime. Because of this curvature, even light is bent by the effects of gravity – an effect we see when we look at some very distant galaxies.

The Earth itself causes a curve in spacetime. Gravity makes light travelling from distant stars bend.

ORBITS & CENTRIPETAL FORCE

the 30-second theory

3-SECOND THRASH

Centripetal force holds a moving body in orbit by pulling it towards the centre – the body is in free fall, but perpendicular motion keeps it up.

3-MINUTE THOUGHT

For a car moving in a clockwise circle the centripetal force is to the right, but the car's driver feels a subjective force to the left – outwards rather than inwards. This apparent centrifugal ('centre-fleeing') force arises because, in the language of relativity, the driver is sitting in a non-inertial frame – one that is rotating rather than moving in a straight line at a constant speed.

If a moving body is subjected to a force in the same direction as its motion, it speeds up. A force applied at right angles, on the other hand, has no effect on speed – only on the direction of motion. If the magnitude of the force remains constant, and always acts perpendicular to the direction of motion, the result is a circular orbit with the force directed towards its centre. Such a force is called centripetal, from the Latin for 'centre-seeking'. The most familiar example of a centripetal force in action is the case of a person whirling a small weight around on the end of a string, but a satellite orbiting the Earth or a planet orbiting the Sun obeys exactly the same principle. In those cases, the centripetal force is provided by gravity. A satellite in orbit is in free fall, but its sideways motion is sufficient to ensure that it always misses the Earth. Because gravitational force depends on distance, there is only one speed at any given radius that results in a circular orbit. If an orbiting body has a different speed to the one appropriate for its altitude, it will follow an elliptical path rather than a circular one. This is still a consequence of centripetal force, but one that varies in strength around the orbit rather than remaining constant.

RELATED TOPICS

See also
FORCE & ACCELERATION
page 78

GRAVITY
page 82

GALILEAN RELATIVITY
page 102

3-SECOND BIOGRAPHIES

CHRISTIAAN HUYGENS
1629–95
Dutch scientist who coined the term centrifugal force and derived the mathematical formula for it

ISAAC NEWTON
1643–1727
Introduced the concept of centripetal as opposed to centrifugal force – and showed how it could explain planetary orbits

30-SECOND TEXT

Andrew May

Speed and orbiting height must be calculated and maintained very precisely to keep a satellite in a circular orbit around the Earth.

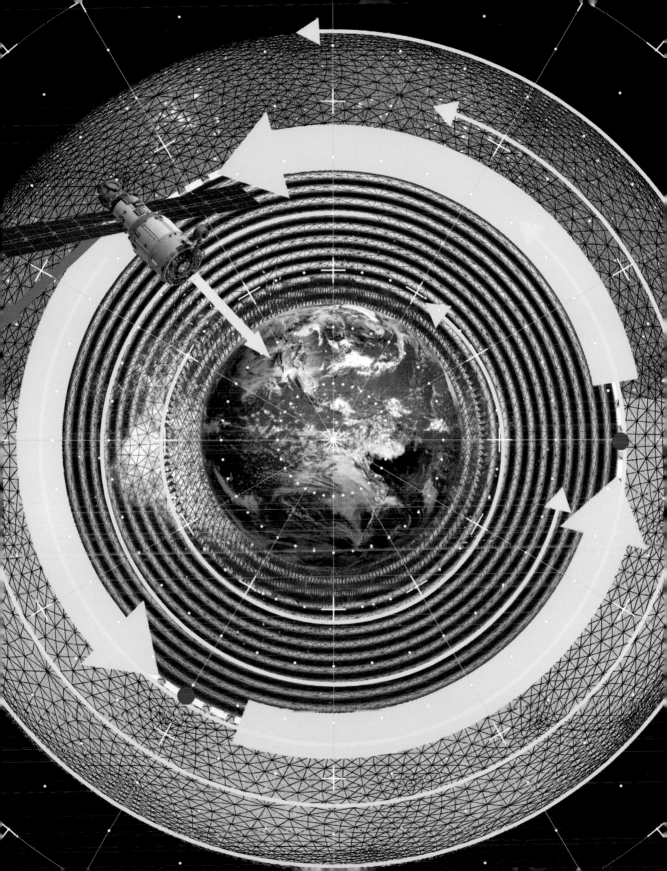

4 January 1643
Born at Woolsthorpe
Manor, Lincolnshire

1661
Becomes a student at
Trinity College,
Cambridge

1665
Returns to Woolsthorpe
for 18 months; starts
thinking about gravity

1667
Appointed a fellow of
Trinity College

1669
Becomes Lucasian
professor of mathematics

1671
His reflecting telescope is
shown to the Royal
Society

1672
Elected a fellow of the
Royal Society

1687
His book, *Philosophiae
Naturalis Principia
Mathematica*, is
published

1696
Moves to London, as
Warden of the Royal Mint

1700
Promoted to Master of
the Mint

1703
Becomes President of the
Royal Society

1704
Publication of *Opticks*

1705
Knighted by Queen Anne

31 March 1727
Dies in Kensington, near
London

ISAAC NEWTON

Isaac Newton was born on a

farm in rural Lincolnshire. He might have been destined for nothing greater than taking over the farm, had it not been for a scholarly uncle and a sympathetic schoolteacher who recognized his academic potential. They secured a place for him at Cambridge University, where he began his studies in 1661. Soon after he graduated, in 1665, the university was closed due to fears of plague, and Newton was forced to return home for an extended vacation. He took his work with him – various scientific problems that had puzzled him as a student – and seized the opportunity to ponder them in depth. Although he published nothing at the time, he laid some of the groundwork for his great discoveries – and experienced the famous incident of the falling apple.

Newton returned to Cambridge in 1667, and soon afterwards set himself the task of constructing a new kind of telescope – using mirrors rather than lenses. This was a technical challenge as well as a scientific one, and the result was so impressive that it came to the attention of the country's leading scientists at the Royal Society in London. They made Newton a fellow of the Society in 1672, and for a short time he was the talk of the scientific community. Unfortunately, however, Newton had an intolerance for criticism, and before long he chose to withdraw from scientific discourse rather than engage in constant argument. It was only in the 1680s that he was drawn back into the fray, largely through the efforts of Edmund Halley who was seeking a mathematical theory of planetary orbits. With Halley's encouragement, Newton succeeded in formulating such a theory using the concept of universal gravitational attraction, which he expounded at length in his book *Philosophiae Naturalis Principia Mathematica* ('Mathematical Principles of Natural Philosophy').

The *Principia* was the first systematic attempt to explain a range of physical phenomena in mathematical terms, and it secured Newton's reputation as the country's foremost scientific genius. In 1696 he was awarded the prestigious post of Warden of the Royal Mint, and he was promoted four years later to the even more lucrative position of Master of the Mint. His second major book, *Opticks*, was published in 1704, although most of the work had been carried out decades earlier. Newton was knighted the following year, 1705, and eventually died in 1727 at the age of 84.

THE WEAK NUCLEAR FORCE

the 30-second theory

One of the four fundamental forces (together with the electromagnetic force, the strong nuclear force and gravity), the weak force is responsible for changing the identities of fundamental particles – as, for example, in radioactive beta decay – and for the conversion of hydrogen to helium in the Sun. Called weak because it is considerably feebler than the electromagnetic force in such cases, its lack of strength is the reason why the Sun barely stays alight. If the weak force had the same strength as the electromagnetic force, the Sun would have burned out long before evolution had time to produce life on Earth. The weak force is transmitted by W or Z bosons, which are about 90 times heavier than a hydrogen atom, but are otherwise similar to the photon of QED. It is the large energy needed to materialize a W or Z boson that enfeebles the weak force at low energy. At very high energies, however, similar to those in the first moments of the big bang, the weak force loses its impotence and merges with the electromagnetic force into a single electroweak force. Neutrinos do not feel the strong or electromagnetic forces, and so are useful probes of the weak force.

In radioactive beta decay a neutron turns into a proton, which remains in the nucleus, and an electron, which leaves – becoming a beta particle.

THE STRONG NUCLEAR FORCE

the 30-second theory

3-SECOND THRASH
Atomic nuclei could not survive were it not for the strong attractive force between neutrons and protons.

3-MINUTE THOUGHT
Electrical forces in atoms are the source of chemical energy and explosives; the strong force in atomic nuclei is the source of nuclear power and nuclear weapons. Grand unified theories postulate that the strong, weak and electromagnetic forces are different aspects of a single unified force, which was present in the aftermath of the big bang and may be manifested at energies higher than are currently accessible experimentally.

One of the four fundamental forces (together with the electromagnetic force, the weak nuclear force and gravity), the strong force binds quarks and/or antiquarks to make hadrons (strongly interacting particles such as protons and neutrons) and binds these together in atomic nuclei. Electrical repulsion among protons in a nucleus would destroy it were it not for the strong attractive force. Neutrons and protons attract one another with the same strength as either attracts its own kind. Within a nucleus, where these constituents are in close proximity, this strong attraction is more than 100 times more powerful than the electrical repulsion. There is a limit, however, to the number of protons that can exist like this: for any individual proton, the attraction only acts between it and immediate neighbours, whereas electrical disruption acts over the entire volume of the group. In a large nucleus this electrical disruption can exceed the localized strong attraction and the nucleus cannot survive. Interplay between the strong attraction and electrical disruption helps to determine the relative stability of different combinations of neutrons and protons. Nuclei seek stability, and one way is to adjust the neutron-proton ratio by beta decay, which is caused by the weak force.

RELATED TOPICS
See also
PHOTONS
page 40

QED
page 68

THE WEAK NUCLEAR FORCE
page 88

3-SECOND BIOGRAPHIES
ERNEST RUTHERFORD
1871–1937
New Zealand-born physicist who effectively discovered the proton

JAMES CHADWICK
1891–1974
English physicist who discovered the neutron

HIDEKI YUKAWA
1907–81
Japanese physicist who predicted that the strong nuclear force arises from the exchange of particles known as pions

30-SECOND TEXT
Frank Close

The strong nuclear force is unleashed by the hydrogen bomb; it also powers the stars.

FIELD OR PARTICLE?

the 30-second theory

The forces of nature are often described using field theory. A field is anything that has values that vary from place to place throughout spacetime. The height above sea level on the Earth provides a two-dimensional picture of a field. Anywhere on the Earth has an altitude. We can draw field lines on a map (contours) that represent locations of equal value. And changing the field value of an object involves a transfer of energy (say, in moving something from the bottom of a hill to the top). Similarly, fundamental forces like electromagnetism can be considered the outcome of a field that varies throughout space and time, with a value at any particular point in this four-dimensional environment. However, this isn't the only approach. It can be useful to consider a force to be the result of an exchange of 'carrier' particles between the matter particles involved. So, for example, electromagnetism can be considered an exchange of photons, the carrier particle of electromagnetism. The same goes for gluons, the massless particles producing the strong force, while W and Z bosons are involved in the weak force. Gravity may involve an equivalent particle, the graviton, though general relativity takes a different, geometric approach to that used in the other forces.

RELATED TOPICS
See also
ELECTROMAGNETISM
page 80

THE WEAK NUCLEAR FORCE
page 88

THE STRONG NUCLEAR FORCE
page 90

3-SECOND BIOGRAPHIES
RICHARD FEYNMAN
1918–88
American physicist whose diagrams demonstrate the role of photons in carrying the electromagnetic force

STEVEN WEINBERG
1933–
American physicist who showed how electromagnetism and the weak nuclear force could be unified

30-SECOND TEXT
Brian Clegg

3-SECOND THRASH
The fundamental forces can be considered the result of variations in a field that extends through spacetime or an interchange of 'force-carrier' particles.

3-MINUTE THOUGHT
Richard Feynman said 'I want to emphasize that light [the carrier of electromagnetism] comes in this form – particles,' while Steven Weinberg commented '[T]he inhabitants of the universe [are] conceived to be a set of fields … and particles [are] reduced to mere epiphenomena.' In practice, both fields and particles are useful to predict the outcomes of the fundamental forces, each being more useful in some circumstances. Both are models – neither is what is 'really' out there.

With field theory physicists can draw a 'map' showing the electromagnetic force; but thinking in terms of particles helps, too.

MOTION

acceleration The rate at which velocity changes. In physics, acceleration covers both increasing velocity (positive acceleration) and decreasing velocity (negative acceleration), which outside of physics is called deceleration. Because velocity is a vector, acceleration can be either or both of a change in speed and a change in direction.

chemical bonds The electromagnetic attraction between atoms (or more specifically between the subatomic particles within the atoms) in a molecule that holds them together. Stronger covalent bonds share electrons between atoms, while weaker ionic bonds are produced by an electromagnetic attraction between a positively charged ion (an atom that has lost one or more electrons) and a negatively charged ion (an atom that has gained one or more electrons).

hydrogen bonding A particular type of electromagnetic attraction between two molecules where a relatively positively charged hydrogen atom is attracted to another atom that is relatively negatively charged. Hydrogen bonding is best known in water, between hydrogen and oxygen. It can have a significant influence on the physical properties of a substance – without hydrogen bonding, water would boil at around -100°C (-148°F).

inertia The tendency of a body with mass to resist changes in its velocity, requiring a force to speed it up or slow it down.

kinetic energy The energy of an object due to its motion. The energy is proportional to the mass of the object and to the square of its velocity. Doubling the velocity quadruples the kinetic energy.

momentum In classical physics, the mass of an object times its velocity. In quantum physics, momentum is Planck's constant divided by the wavelength associated with the quantum particle, which applies to both massive particles and those without mass, like a photon.

Newton's first law of motion Also called the law of inertia. It says that a body will stay at rest or in uniform motion in a straight line (that is, with a fixed velocity) unless a force acts on it.

Newton's second law of motion Originally in the form that a change of motion is proportional to the force applied and takes place in the direction of the application of force, it is now simply stated as $F=ma$, where F is the force applied, m is the mass of the object the force is applied to and a is the resultant acceleration – the rate of change of the object's velocity.

Newton's third law of motion Usually stated as 'Every action has an equal and opposite reaction.' The result is that if you push something, it pushes back on you, as demonstrated in the recoil of a gun or a rocket motor in flight, where a force backwards on the fuel produces an opposite force forwards on the rocket. This is why a rocket can fly in a vacuum with nothing external to push against.

scalar A property that only has a numerical value, like mass, is a scalar (as opposed to a vector).

Van der Waals force Electromagnetic attraction or repulsion between molecules, excluding the stronger forces due to chemical bonds and hydrogen bonding.

vector A property that has a numerical value and a direction is a vector. Velocity, for instance, is made up of speed (a scalar) and the direction of that speed.

velocity A vector, made up of the speed at which something is moving and the direction in which it is moving.

MOVEMENT, SPEED & VELOCITY

the 30-second theory

RELATED TOPICS
See also
FORCE & ACCELERATION
page 78

GALILEAN RELATIVITY
page 102

NEWTON'S LAWS
page 104

3-SECOND BIOGRAPHIES
ZENO OF ELEA
C.490–C.430 BCE
Greek philosopher who proposed various paradoxes suggesting movement is impossible

GALILEO GALILEI
1564–1642
Italian natural philosopher who conducted experiments on the motion of bodies

ISAAC NEWTON
1643–1727
English physicist who codified the principles of dynamics in his three laws of motion

30-SECOND TEXT
Andrew May

3-SECOND THRASH
Speed measures the rate at which an object moves; velocity measures not just the rate of movement but its direction as well.

3-MINUTE THOUGHT
In order to analyse movement, modern physicists use the mathematical techniques of calculus and differential equations, which divide motion up into infinitesimally small chunks. Before such methods were invented, the subject confused even the world's greatest thinkers. The Greek philosopher Zeno, who belonged to a school of thought that held change of any kind to be an illusion, formulated a number of paradoxes that appear to show that movement is impossible.

The speed of an object is a numerical measure of its rate of movement relative to some other point that is arbitrarily taken as 'fixed'. Speed is calculated by dividing the distance travelled by the time taken. The result is expressed in length units per time unit; for example metres per second or miles per hour. In mathematical terms speed is a scalar quantity, represented by a single number. Velocity, on the other hand, is defined as a vector quantity having both magnitude and direction. The magnitude of the velocity is simply the speed, while its direction is the direction of motion at that particular point in time. The use of vectors instead of scalars becomes important in the science of dynamics, which deals with the way an object's motion changes in response to applied forces. Like velocity, forces are vector quantities. If a force is applied in the same direction as the velocity, the result is an increase in speed and no change in direction. If the force is applied at an angle, however, it produces a change in direction as well as speed. When Newton's second law tells us that acceleration is proportional to applied force, 'acceleration' means overall change in velocity – direction as well as speed.

Speed is distance per time unit, for example km/h, but velocity measures direction, as well.

MOMENTUM & INERTIA

the 30-second theory

3-SECOND THRASH
Momentum is mass
times velocity, and the
total momentum of a
closed system never
changes: an isolated object
will remain in the same
state of motion.

3-MINUTE THOUGHT
The mass that appears in
the formula for momentum
is referred to as 'inertial
mass', because it is what
gives an object its inertia.
In classical physics, this
is distinct from the
'gravitational mass'
appearing in Newton's
formula for the force of
gravity. Nevertheless,
the two quantities appear
to be identical to each
other, and no experiment
has been able to
distinguish between them.

The momentum of an object is defined as its velocity multiplied by its mass. Momentum is a conserved quantity; in other words it always remains constant within a closed system. When a number of objects interact with each other, there may be an exchange of momentum between them but the total momentum always stays the same. Due to the conservation of momentum, an object that does not interact with its surroundings will always retain its current state of motion. If it is stationary it will remain stationary; if it is moving it will continue moving at a constant velocity. This resistance to change is called the principle of inertia, and is the basis for Newton's first law of motion. His second law states that when an external force is applied to the object, the rate at which its momentum changes is equal to the applied force. Because momentum is the product of mass and velocity, this means the force needed to change the velocity by a given amount is proportional to the object's mass. In other words the more massive an object is, the more it feels the effects of inertia.

RELATED TOPICS
See also
MASS
page 18

FORCE & ACCELERATION
page 78

GRAVITY
page 82

3-SECOND BIOGRAPHIES
GALILEO GALILEI
1564–1642
Italian natural philosopher
who discovered the principle
of inertia

RENÉ DESCARTES
1596–1650
French philosopher who
proposed the law of
conservation of momentum
in an early form

ISAAC NEWTON
1643–1727
English physicist who codified
the principles of inertia and
momentum

30-SECOND TEXT
Andrew May

*The cue ball transfers
its momentum to the
other pool balls,
breaking them apart.*

GALILEAN RELATIVITY

the 30-second theory

3-SECOND THRASH
All motion is relative –
when you're sitting in an
armchair you are actually
whizzing through space at
more than 700,000
kilometres per hour!

3-MINUTE THOUGHT
In Galilean relativity we
simply add speeds, so if
you are sitting in a railway
carriage moving at 100
kilometres per hour and
you roll a ball along the
carriage at 10 km/h,
someone on the platform
would measure the ball's
speed to be 110 km/h.
However, this is only true
as long as we are travelling
slowly compared to the
speed of light; as we
approach the speed of light
we cannot simply add
speeds. In fact, Galilean
relativity is the low-speed
approximation of Einstein's
special relativity.

When Galileo argued that the
Earth orbited the Sun, one of his opponents'
main objections was that we don't feel the Earth
moving. He thought about this, and realized that
all motion is relative. If we are moving at a
constant speed in a straight line, he said, there
are no mechanical experiments that could
determine whether we are moving or at rest.
An object dropped from the top of the mast of
a boat sailing on a perfectly smooth lake will hit
the deck at the bottom of the mast, just as if the
boat were at rest. A pendulum swinging back
and forth on the boat will swing at the same rate
whether the boat is moving or at rest. When
we are sitting on a train or an aeroplane, so long
as there are no accelerations a cup of water in
front of us will have a perfectly flat surface and
objects will not move about. In fact, of course,
when we are sitting in our armchairs, the Earth is
orbiting the Sun at about 1,073 km/h (667 mph)
and the Sun is hurtling around the centre of the
Milky Way galaxy at roughly 708,000 km/h
(440,000 mph).

RELATED TOPICS
See also
SPEED OF LIGHT
page 50

SPECIAL RELATIVITY
page 112

3-SECOND BIOGRAPHIES
GALILEO GALILEI
1564–1642
Italian natural philosopher, the
first person to realize that all
motion is relative

ALBERT MICHELSON
1852–1931
American physicist who tried
to measure the Earth's motion
through the ether

ALBERT EINSTEIN
1879–1955
German-born physicist
who generalized Galilean
relativity to include
experiments with light

30-SECOND TEXT
Rhodri Evans

*Juggling is possible
because you feel like
you're sitting still even
though you (and the
Earth) are zooming
through space.*

NEWTON'S LAWS

the 30-second theory

Newton's three laws of motion first appeared in print in 1687, at the start of his famous book *Philosophiae Naturalis Principia Mathematica* ('Mathematical Principles of Natural Philosophy'). The first law simply states the principle of inertia: that an object will remain at rest, or moving at constant velocity, unless it is acted on by an external force. The second law goes on to describe how the motion of an object is altered when a force is applied to it: the change in its momentum, per unit time, is equal to the applied force. Since momentum is mass multiplied by velocity, Newton's second law implies that for an object of constant mass m, the applied force F and the resultant acceleration a are related by the famous equation $F=ma$. As for the third law, it is commonly expressed in the form 'for every action there is an equal and opposite reaction'. In other words, if object A exerts a force on object B, then object B exerts exactly the same force on object A, but in the opposite direction. All three of these laws can ultimately be viewed as consequences of the conservation of momentum.

3-SECOND THRASH
Applying a force to an object changes its motion, while at the same time the object itself exerts an equal force in the opposite direction.

3-MINUTE THOUGHT
As simple as they appear on the surface, Newton's laws of motion are literally rocket science. A spaceship that is a long way from a gravitational field will cruise indefinitely at constant velocity, as long as its rocket engine is switched off. When the engine is switched on, it pushes the spaceship forwards with added momentum – but only because the rocket exhaust is shooting out of the back with equal and opposite momentum.

RELATED TOPICS
See also
FORCE & ACCELERATION
page 78

MOVEMENT, SPEED & VELOCITY
page 98

MOMENTUM & INERTIA
page 100

3-SECOND BIOGRAPHIES
GALILEO GALILEI
1564–1642
Italian scientist who wrote precursors to Newton's laws

ISAAC NEWTON
1643–1727
English physicist who published his three laws of motion in 1687

EDMUND HALLEY
1656–1742
English astronomer and mathematician who personally published Newton's *Principia*

30-SECOND TEXT
Andrew May

The momentum added by the rocket engine is equal and opposite to the momentum of the exhaust.

FRICTION

the 30-second theory

Thank goodness for friction.

Without this force that resists the sliding of one surface or layer past another, buildings would fall down and swimming, driving and even walking would be impossible. On the other hand friction is a nuisance, too: it makes machinery inefficient, slows down ships and airplanes, and ultimately causes moving parts to wear down and fail, from cogs to knee joints. Friction is caused largely by the attractive forces that exist between objects in close contact, in particular the so-called van der Waals or dispersion forces that arise from interactions between floppy clouds of electrons in atoms and molecules. These operate only over very short distances of a few nanometres, and they are weaker than chemical bonds. But over a large contact area they can add up to a significant force, and the gecko's adhesive feet exploit this. As well as these static forces, friction arises dynamically from relative motion. Here, too, the forces are caused partly by interatomic attractions, but dynamic friction can also result from interlocking or collisions of tiny bumps (asperities) on the surfaces. Friction converts some kinetic energy of sliding into heat, which is why rubbed hands get warmer. Both asperity collisions and heat dissipation can damage surfaces sliding over one another.

RELATED TOPICS
See also
FORCE & ACCELERATION
page 78

KINETIC ENERGY
page 122

HEAT
page 138

3-SECOND BIOGRAPHIES
GUILLAUME AMONTONS
1663–1705
French inventor who discovered the laws of dry friction

DAVID TABOR
1913–2005
English physicist who initiated the modern field of tribology (the study of friction)

30-SECOND TEXT
Philip Ball

3-SECOND THRASH
Friction is a force that resists the relative motion of surfaces and ultimately changes kinetic energy to heat.

3-MINUTE THOUGHT
Surprisingly, dry friction doesn't depend on the areas of the surfaces in contact. That's because there's rather little real contact even for apparently smooth surfaces: microscopic asperities prop them apart, so they touch in only a few places. As the contact gets more intimate, friction increases – which is how putty-like adhesives hold things on the wall, by penetrating under pressure into tiny cracks and valleys. So do the hairs of a gecko's feet.

Thanks to weak electrical interactions between all objects at very short distances, their relative movement is opposed by friction.

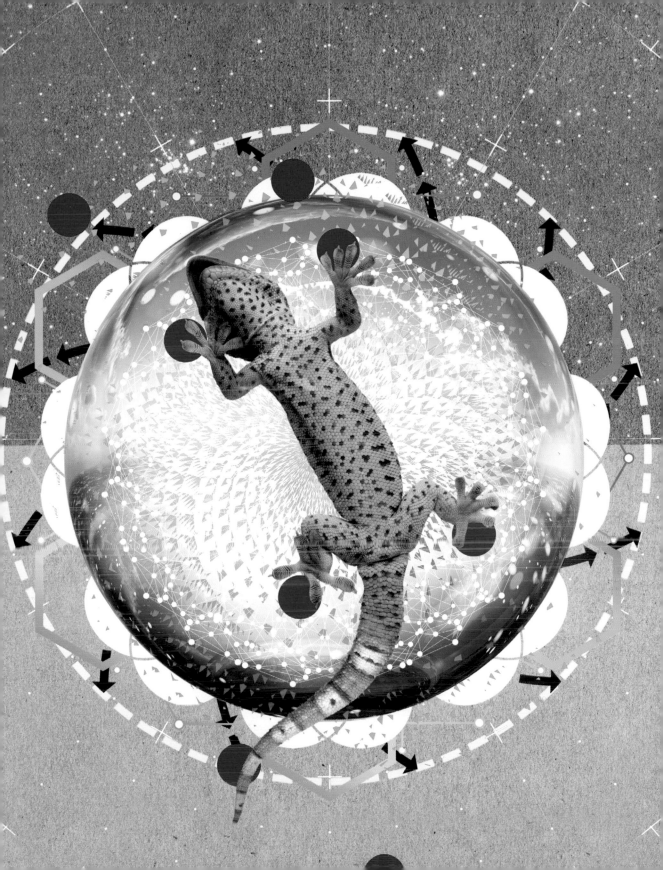

14 March 1879
Born in Ulm, Germany

1885
Starts school in Munich

1895
Leaves Germany for Switzerland

1900
Gains his first degree from Zurich Polytechnic

1902
Starts work at the Swiss Patent Office

1905
Publishes several major scientific papers, including one on relativity

1906
Awarded a doctorate by Zurich University

1909
Becomes associate professor of physics at Zurich University

1911
Moves to Prague University

1912
Returns to Zurich as a professor at the Polytechnic

1914
Obtains a permanent professorship in Berlin

1915
Publishes general theory of relativity

1919
Relativity becomes front-page news following Arthur Eddington's eclipse expedition

1921
Wins the Nobel Prize for physics

1933
Moves to the Institute for Advanced Study in Princeton, New Jersey

1939
Warns President Roosevelt about the military potential of atomic weapons

18 April 1955
Dies in Princeton, New Jersey

ALBERT EINSTEIN

In 1880, when Albert Einstein was a year old, his family moved to Munich, where his father and uncle set up an electrical business. Throughout his childhood Einstein was a voracious learner, but often hated the way subjects were taught at school. In 1894, when the family business moved to Italy, 15-year-old Albert was left alone and unhappy in Munich. Within a year he had quit school, renounced his German citizenship and moved to Switzerland. His heart was set on studying physics at Zurich Polytechnic ... and in 1896, at the unusually young age of 17, he passed the entrance examination. Unfortunately, he was every bit as infuriated by the style of teaching at the Polytechnic as he had been at school, and so he regularly found himself in conflict with the staff. That may be the reason why, after Einstein graduated in 1900, he found it impossible to obtain an academic job.

The best position he could find, after almost two years of searching, was 'technical expert, third class' at the Swiss Patent Office in Bern. He stayed there for seven years and, astonishingly, some of his greatest scientific work was done during this period. There was ample opportunity, while he was sitting at his desk waiting for patent applications to come in, to engage in profound theoretical thinking about the major scientific problems of the day. In the course of just one year, 1905, he published no fewer than four groundbreaking papers – on quantum theory, molecular dynamics, on relativity ... and introducing his most famous equation: $E=mc^2$. The papers were so revolutionary, in fact, that the scientific community was slow to recognize their significance. Only in 1909 did he obtain a full-time academic position, at the University of Zurich. This was followed by a succession of increasingly prestigious posts, culminating in 1914 in a full professorship.

The following year Einstein put the finishing touches on his masterpiece, the general theory of relativity – a new theory of gravity, replacing that of Newton. One of the novel predictions made by Einstein's theory was confirmed by an English astronomer, Arthur Eddington, during the solar eclipse of 1919. This event propelled Einstein to the status of an international celebrity, which he retained for the rest of his life. He made several visits to the United States, and in 1933 – with the Nazis making Germany a singularly unpleasant place to live – he moved there permanently. He accepted a position at the recently formed Institute for Advanced Study in Princeton and remained there until his death at the age of 76.

FLUID DYNAMICS

the 30-second theory

The circulation of the oceans and atmosphere, the flow of water down a pipe, the swirling of smoke in the air and the churning of liquid iron in the Earth's core: all are described by the theory of fluid dynamics, also known (because of its long association with water) as hydrodynamics. It is renowned for being one of the hardest problems in science – not because the basic physics is hard to understand, but because the equations are usually so tricky to solve. These equations, called the Navier-Stokes equations, were first written down in the 19th century. They apply Newton's second law of motion to all parts of the fluid, describing how it moves according to the forces acting on it: how the velocity, pressure, temperature and density are related at all points in the fluid. They can be solved for some particularly simple kinds of flow, but generally this is too difficult to do with pen and paper because all parts of the fluid affect each other. The equations are typically solved numerically by computer: making a guess at the correct pattern of flow and then refining it. Fluid flow is particularly complex when it is turbulent, which happens, for example, if the flow is driven hard. Richard Feynman called turbulence 'the most important unsolved problem of classical physics'.

RELATED TOPICS
See also
LIQUIDS
page 22

FORCE & ACCELERATION
page 78

NEWTON'S LAWS
page 104

3-SECOND BIOGRAPHIES
DANIEL BERNOULLI
1700–82
Swiss mathematician who wrote one of the first books on hydrodynamics

GEORGE GABRIEL STOKES
1819–1903
Irish mathematical physicist who helped establish the basic laws of fluid motion

OSBORNE REYNOLDS
1842–1912
Northern Irish physicist who explained the transition from smooth to turbulent flow

30-SECOND TEXT
Philip Ball

Daunting equations describe movements from the Earth's core to the interactions of oceans and atmosphere.

3-SECOND THRASH
Fluid dynamics, the theory that describes how fluids move and flow, is derived from the fundamental laws of motion.

3-MINUTE THOUGHT
When fluid flows are turbulent, their movements are typically chaotic, meaning they become impossible to predict beyond a certain point in time. Tiny disturbances at one time, too small to measure, might bloom to change the entire flow pattern in the future. This is why weather prediction becomes impossible more than about ten days ahead: no matter how good our data and computers are, chaos renders the weather unknowable beyond that time.

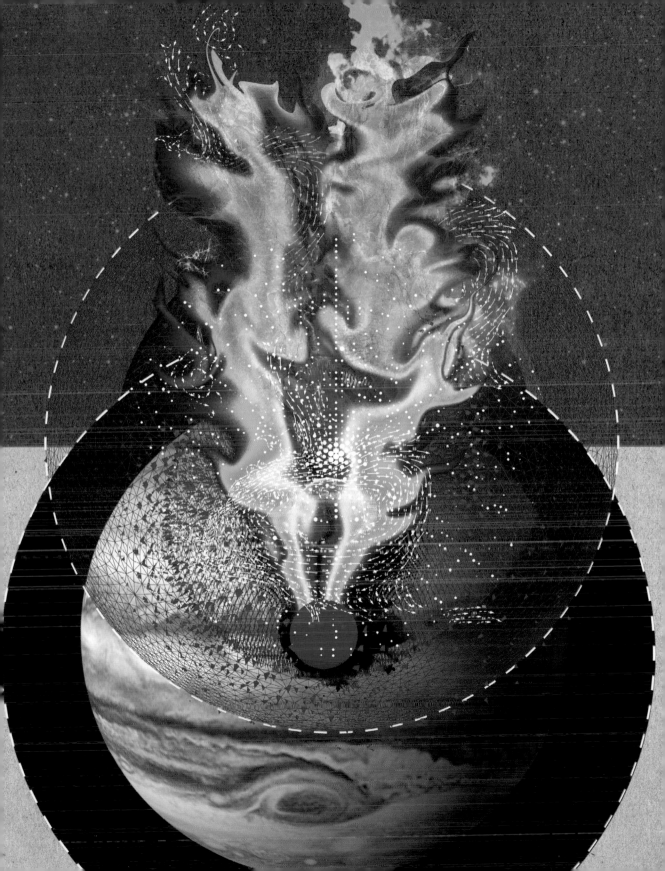

SPECIAL RELATIVITY

the 30-second theory

3-SECOND THRASH
Einstein argued that the speed of light is the same for everyone; this means that lengths get shorter and time gets dilated as we approach the speed of light.

3-MINUTE THOUGHT
Because of time dilation, it would be possible in theory to leave your twin on Earth, go on a space flight you think lasts, say, five years and return to find that your twin is 20 years older! We see the effect of time dilation everyday in particle accelerators, but in our own lives we travel at such a tiny fraction of the speed of light that the effects of special relativity are negligible.

Galileo's original version of relativity says that in an enclosed space without windows it would be impossible to distinguish steady, non-accelerating movement from stillness. In the late 1800s, with the development of electromagnetism, some physicists suggested that an experiment involving light would prove Galileo wrong. This possibility troubled Albert Einstein, and in 1905 he wrote a landmark paper bringing light into the relativistic picture. Light depends on moving at a particular speed to support its interplay of electricity and magnetism: Einstein suggested this meant that the speed of light in a vacuum stays the same, however fast you move towards or away from it. The consequences of these two suggestions are far-reaching – time and space are no longer absolute. Observers moving at different speeds will measure different lengths for a ruler, while 1 second of time will seem different, too: it depends on how quickly one is moving. Time will run more slowly if you approach the speed of light. This theory also led to possibly the most famous equation in physics: $E=mc^2$ (E is the energy, m is the mass and c, the speed of light). This tells us that mass is a concentrated form of energy. Special relativity also tells us that the speed of light is a cosmic speed limit: nothing can travel faster than it.

RELATED TOPICS
See also
SPEED OF LIGHT
page 52

ELECTROMAGNETISM
page 80

GALILEAN RELATIVITY
page 102

3-SECOND BIOGRAPHIES
KENDRICK LORENTZ
1853–1928
Dutch theoretical physicist whose transformation equations are part of the foundation of special relativity

HENRI POINCARÉ
1854–1912
French mathematician, theoretical physicist and philosopher

ALBERT EINSTEIN
1879–1955
German-born theoretical physicist who revolutionized our understanding of space, time and gravity

30-SECOND TEXT
Rhodri Evans

With Einstein's famous equation we see that mass is a form of energy.

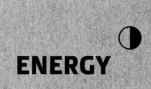

ENERGY

binding energy The energy required to hold particles together. In the case of the atomic nucleus, the binding energy is provided by the strong force. With light atoms, the energy required to bind the nucleus decreases as the nucleus gets bigger, so energy is released when extra particles are bound into the nucleus: this is nuclear fusion. With atoms that are heavier than iron, extra energy is required to hold them together, so that when an atom splits apart to form smaller atoms energy is released: this is nuclear fission.

chemical bonding energy The energy in the bonds that link together atoms to produce molecules. In a chemical reaction, if the total bonding energy of the initial molecules is greater than the total bonding energy of the product molecules, the reaction will give off heat: this is how most biological processes and burning are powered. In some reactions, the resultant molecules have greater chemical bonding energy than the initial molecules and the result is a reaction that takes in heat to make it work.

conservation of momentum A number of physical properties, such as energy, are conserved – that is, stay constant – in a closed system (a system that has no connection with the outside universe). One of these properties is momentum – classically, the mass of an object times its velocity. In quantum physics, momentum is Planck's constant divided by the wavelength associated with the quantum particle, which applies to both massive particles and those without mass, like a photon.

dispersion forces The weakest of the possible electromagnetic forces between atoms in different molecules, caused by the electrons in an atom being more to one side of the atom than the other, giving it a slight negative charge on that side and a slight positive charge on the other.

grand unified theory A physical theory that combines three of the four forces of nature: the electromagnetic, strong and weak forces. While there are some theories that attempt to add in gravity as well, like string theory (making this a 'theory of everything'), they are highly speculative and as yet do not make testable predictions.

interatomic attractions Electromagnetic attraction between separate atoms (or the atoms in separate molecules), such as dispersion force, van der Waals force and hydrogen bonding. These attractions make it harder to separate the linked atoms or molecules, producing changes to the physical properties of a substance like increased boiling point.

kinetic energy The energy of an object due to its motion. The energy is proportional to the mass of the object and to the square of its velocity. Doubling the velocity quadruples the kinetic energy.

nanoscale Objects and processes operating at around a nanometre (a billionth of a metre) in size – the scale of operation of nanotechnology.

potential energy The energy due to the state of a system – for example, the gravitational energy available when an object is lifted up to a high place and can then be dropped, or the energy that is stored in chemical bonds.

power The rate at which work is performed – the amount of energy used per second.

proton A positively charged quantum particle, most frequently found in the nucleus of an atom. Protons are composed of three fundamental particles: two up quarks and one down quark. Although repelled by other protons, because like electrical charges repel, when very close the strong nuclear force that holds the quarks together becomes stronger than the electromagnetic repulsion, making the nucleus stable. The number of protons in an atom determine which element the atom is – the 'atomic number' of an element is the number of protons it has. A single proton makes up the nucleus of the most basic atom, hydrogen.

vector A property that has a numerical value and a direction is a vector. Velocity, for instance, is made up of speed (a scalar) and the direction of that speed.

WORK & ENERGY

the 30-second theory

Energy is one of those terms that most of us use freely without being exactly sure what we are talking about. Special relativity showed that mass and energy were interchangeable, but for practical purposes energy is the phenomenon that makes things happen – that drives change. Unlike a force, which is a vector (having size and direction), energy is a scalar (simply a quantity). The amount of energy is measured in joules, although the old unit of energy, the calorie, still crops up in one form of chemical energy: the energy content of food. (Confusingly, when 'calorie' is used on a food packet, it actually means kilocalorie, but nutritionists thought the 'kilo' part would confuse the public.) Although energy can come from a variety of sources, it is usually of interest because it can do work. In essence, work is simply energy transferred from one place or form to another. So, for instance, when we use up chemical energy from our body to give a car kinetic energy as we push it along a road (and perhaps potential energy if it is going up a hill), we are doing work in the sense used in physics.

RELATED TOPICS
See also
SPECIAL RELATIVITY
page 112

POWER
page 120

MACHINES
page 132

3-SECOND BIOGRAPHIES
WILLIAM GROVE
1811–96
Welsh physicist who first suggested the equivalence of different forms of energy

JAMES JOULE
1818–89
English physicist who formulated the relation of heat to mechanical work

EMMY NOETHER
1882–1935
German mathematician who proved the link between symmetries in a system and conservation laws

30-SECOND TEXT
Brian Clegg

Pushing uphill uses your body's chemical energy to give the car kinetic and potential energy.

3-SECOND THRASH
Energy is the driver for making things happen and for change, while work is the transfer of energy from one location or form to another.

3-MINUTE THOUGHT
Energy, or more precisely mass/energy, is conserved in a closed system: this is the first law of thermodynamics. This means that you can't produce work from nowhere. While the conservation of energy was originally a matter of common sense, it emerges from the closed system being invariant in time. Quantum physics stretches the conservation, allowing mass/energy to come into existence as long as it is for a brief time interval.

POWER

the 30-second theory

Of all the terms in physics that are misused in everyday English, 'power' probably suffers the most. It is often regarded simply as a variant on 'energy', as a loose description of the degree to which something or someone can achieve things. But in physics, power is specifically the rate at which work is performed. Work is a transfer of energy, and power the rate at which energy is transferred – whether down an electrical cable or by a motor. Power is measured in joules per second, known for simplicity as watts. (The power companies' favourite unit, the kilowatt hour, is a clumsy way of measuring energy in lumps of 3,600,000 joules.) On the mechanical front, energy transferred is force times distance moved by the object to which force is being applied. Power is therefore force times distance divided by time – and because distance over time is velocity, power becomes force times velocity. A 'horsepower', originally introduced by James Watt as a means of comparing horse and steam power, and now used to measure the power of engines, is around 0.75 kilowatts. The usual figure given is 'brake horsepower', a nominal unloaded output from the engine, which will be considerably more than the power that can be used.

3-MINUTE THOUGHT
The distinction between the amount of energy in a fuel and the power it produces decides what the fuel will be useful for. Petrol, for instance, releases around 15 times as much energy per kilogram as the explosive TNT. But TNT releases the energy over a much shorter period of time. As power is energy divided by time, the power generated by the TNT is greater, producing greater explosive impact.

RELATED TOPICS
See also
NEWTON'S LAWS
page 104

WORK & ENERGY
page 118

MACHINES
page 132

HEAT ENGINES
page 140

3-SECOND BIOGRAPHIES
JAMES WATT
1736–1819
Scottish engineer after whom the unit of power is named

MICHAEL FARADAY
1791–1867
English scientist who laid the foundations for electric motors

KARL BENZ
1844–1929
German engineer, arguably the first to apply the internal combustion engine to a car

30-SECOND TEXT
Brian Clegg

'Horsepower', a measure of power, comes from steam pioneer James Watt.

KINETIC ENERGY

the 30-second theory

3-SECOND THRASH
Kinetic energy is the energy of motion: the faster an object is moving, the more kinetic energy it will have.

3-MINUTE THOUGHT
At an atomic level, kinetic energy is related to the temperature of an object. The hotter an object is, the more quickly its atoms or molecules are moving. At a temperature of absolute zero all motion would stop, but because this would violate Heisenberg's uncertainty principle, which says we can't know both position and momentum exactly, achieving absolute zero is impossible. Scientists have, however, cooled objects to within a few thousandths of a degree of absolute zero.

Kinetic energy is the energy an object has due to its motion. All moving objects have kinetic energy, including large objects such as stars and planets and tiny objects such as molecules and atoms. An object's kinetic energy depends on its mass and on the square of its speed. If we double an object's mass, but keep the speed constant, its kinetic energy will also double; but if we keep the mass constant and double its speed the kinetic energy will increase by a factor of four. Therefore, a car travelling at 65 km/h (40 mph) has nearly twice the kinetic energy of a car travelling at 50 km/h (30 mph), which is one of the reasons that slowing down in built-up areas is so important. Energy is usually transferred from one form to another, so when an object gains kinetic energy this gain must come from somewhere. For example, in a car it is the explosion of the petrol in the engine's cylinders that is converted to kinetic energy of the pistons and this drives the car forward. When a car brakes to slow down, the loss in the car's kinetic energy is mainly converted to heat in the brake pads and discs.

RELATED TOPICS
See also
THE UNCERTAINTY PRINCIPLE
page 64

WORK & ENERGY
page 118

POTENTIAL ENERGY
page 124

3-SECOND BIOGRAPHIES
WILLEM'S GRAVESANDE
1688-1742
Dutch scientist who showed kinetic energy depended on the square of the speed of an object

WILLIAM THOMSON
1824–1907
Northern Irish physicist, coined the term 'kinetic energy'

LUDWIG BOLTZMANN
1844–1906
Austrian physicist who, together with James Clerk Maxwell, developed the kinetic theory of gases

30-SECOND TEXT
Rhodri Evans

Molecular motion stops at absolute zero – but in practice this state cannot be reached.

POTENTIAL ENERGY

the 30-second theory

Potential energy is the stored energy that an object possesses, either due to its position or its internal properties. There are many types of potential energy including chemical potential energy, the potential energy of a spring and gravitational potential energy. A battery has chemical potential energy, which can be converted to electrical energy when we attach a circuit to the battery's terminals. An object at the top of a building has gravitational potential energy, which is converted to kinetic energy if the object falls to the ground. As the object falls it gains speed and therefore gains kinetic energy; this gain in kinetic energy equals the gravitational potential energy that it loses. When we wind up a clock we store energy in its spring; the spring then slowly releases this stored energy to drive the clock's mechanism and move the clock's hands. A pendulum converts energy back and forth between gravitational potential energy and kinetic energy. The pendulum bob's kinetic energy is zero at each end of its swing, when its potential energy is at its maximum; at the middle of the swing its kinetic energy is at its maximum and its potential energy is at its minimum.

RELATED TOPICS
See also
WORK & ENERGY
page 118

KINETIC ENERGY
page 122

NUCLEAR ENERGY
page 128

3-SECOND BIOGRAPHIES
ALESSANDRO VOLTA
1745–1827
Italian physicist who invented the electric battery

WILLIAM RANKINE
1820–72
Scottish engineer who introduced the concept of potential energy to physics

30-SECOND TEXT
Rhodri Evans

3-SECOND THRASH
Potential energy is the energy an object has stored due to its position or its chemical or physical structure.

3-MINUTE THOUGHT
The nuclei of atoms possess nuclear potential energy. Changes in the structure or the composition of the atoms' nuclei lead to a release of energy, either as radioactivity or in the form of heat and light. The energy stored in this form is much greater per kilogram than the energy stored as chemical potential energy.

As a pendulum swings, it converts potential energy to kinetic energy and back again.

CHEMICAL ENERGY

the 30-second theory

3-SECOND THRASH
Chemical energy is released and absorbed during chemical reactions by making and breaking bonds between atoms; these bonds store energy and they are made by electrons that bind atoms together into molecules.

3-MINUTE THOUGHT
Humans run on chemical energy. The food we eat contains carbohydrates, fats and proteins, which are complex molecules containing carbon and hydrogen atoms. The molecules that make up these foods contain chemical energy locked away in the bonds between their constituent atoms. This energy is released in chemical reactions that take place within the cells of our body. These reactions – called respiration –provide energy to move muscles, work the brain and maintain our metabolism.

Chemical energy powers our world. Coal and gas electricity generation stations and internal combustion engines in cars all rely on releasing energy through combustion. This is a chemical reaction between hydrocarbon molecules in a fuel – for example, wood, coal, gas, petrol or oil – and oxygen molecules in the atmosphere. This produces a combination of carbon dioxide gas and water vapour as well as energy in the form of heat and light. Chemical energy is released during the reaction by breaking and making bonds between atoms. There is more energy in the bonds between the atoms of carbon and hydrogen in hydrocarbon molecules and between the oxygen atoms that make up oxygen molecules than there is between these same atoms when they are rearranged into water molecules and carbon dioxide molecules. This difference in chemical bonding energy is released as heat and light during combustion. Many other different reactions involving a vast number of chemicals are possible. Some reactions release chemical energy and others absorb it. Explosives used in bombs, bullets and fireworks release their chemical energy rapidly. Plants absorb energy from sunlight to power the conversion of carbon dioxide and water into the complex molecules of life, which store this added energy in their chemical bonds.

RELATED TOPICS
See also
WORK & ENERGY
page 118

KINETIC ENERGY
page 122

3-SECOND BIOGRAPHIES
ANTOINE LAVOISIER
1743–94
French chemist who saw that combustion is a chemical reaction that requires oxygen

JOSIAH WILLARD GIBBS
1839–1903
American physicist; discovered that energy stored within chemicals powers reactions

GILBERT NEWTON LEWIS
1875–1946
American chemist whose theories of electron bonding underpin our understanding of chemical energy

30-SECOND TEXT
Leon Clifford

Fuel reaction. Hydrocarbons combine with oxygen. Heat, light, water vapour and CO_2 are the result.

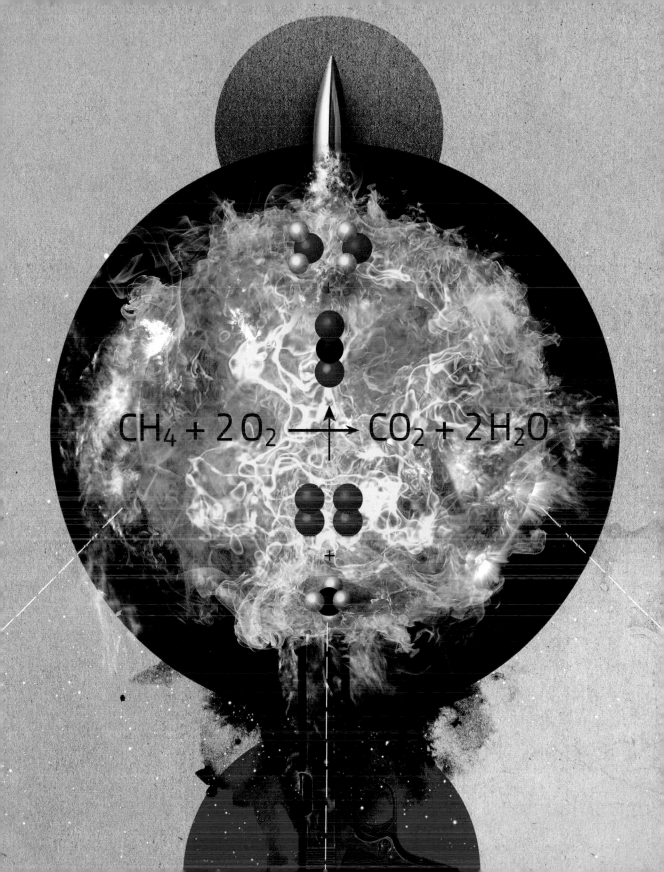

NUCLEAR ENERGY

the 30-second theory

RELATED TOPICS
See also
THE STRONG NUCLEAR FORCE
page 90

WORK & ENERGY
page 118

KINETIC ENERGY
page 122

Nuclear energy – which powers the Sun and stars and warms the Earth's interior – is released from the nuclei at the heart of atoms. Nuclei are composed of protons and neutrons – except for the hydrogen nucleus, which consists of a single proton. Elements with atoms that are heavier than hydrogen atoms have more than one proton and these positively charged particles repel each other. This electrical repulsion is overcome by a force that binds protons and neutrons; energy associated with this binding force is stored within the nucleus. The amount of this stored binding energy depends on the size of the nucleus. When the atomic nuclei of lighter elements are combined in nuclear fusion reactions (found in stars and hydrogen bombs), some of this binding energy is released since not all of it is needed by the larger combined nuclei. However, this is not the case for the atomic nuclei of elements heavier than iron, such as uranium. These release energy when they split apart rather than when they fuse together. This is called nuclear fission and it is found in nuclear reactors and in the radioactive decay that warms the interior of the Earth. Both fusion and fission release excess binding energy from the nuclei of atoms – and this is the source of nuclear energy.

3-SECOND THRASH
Nuclear energy is released when the nuclei of atoms lighter than iron fuse together or the nuclei of atoms heavier than iron break apart.

3-MINUTE THOUGHT
We are the products of nuclear fusion. The carbon and oxygen and all the trace elements that make up our bodies were created in the nuclear furnace at the heart of giant stars that exploded billions of years ago. A sequence of nuclear fusion reactions starting with hydrogen created progressively heavier elements: beryllium, lithium, carbon, nitrogen and oxygen ... So each of us is made from stardust.

3-SECOND BIOGRAPHIES
ALBERT EINSTEIN
1879–1955
German-born physicist whose $E=mc^2$ equation calculates the energy released from nuclear reactions

ARTHUR EDDINGTON
1882–1944
English astronomer who proposed fusion as the mechanism of stars

30-SECOND TEXT
Leon Clifford

Nuclear fission. Uranium-235 gains a neutron, then uranium-236 splits into Krypton and Barium, releasing energy.

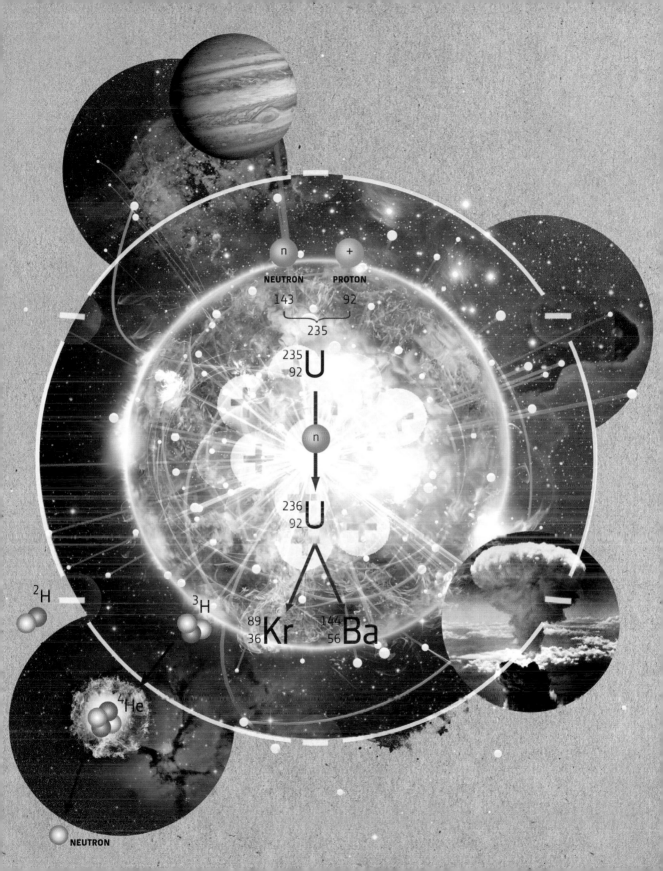

30 August 1871
Born in Spring Grove
(now Brightwater), New
Zealand

1886
Wins a scholarship to the
prestigious Nelson
Collegiate School

1892
Obtains a BA from
Christchurch College,
Canterbury

1893
Obtains an MA with
first-class honours in
physical science

1894
Wins an '1851 Exhibition
Scholarship' to study at
Cambridge University in
England

1898
Appointed professor
at McGill University,
Montreal, Canada

1898–1903
Does important work on
radioactivity, identifying
alpha and beta decay, and
naming the third type
'gamma decay'

1908
Appointed professor at
Manchester University,
England

1908
Awarded the Nobel Prize
for chemistry

1911
Discovers the atomic
nucleus

1917
Bombards nitrogen atoms
with alpha particles,
changing nitrogen to
oxygen; identifies a
positive particle in the
nucleus

1919
Returns to Cambridge to
become director of the
Cavendish Laboratories

1920
Names the positive
particle in the nucleus
the 'proton'

19 October 1937
Dies in Cambridge,
aged 67

ERNEST RUTHERFORD

Rutherford was born in 1871 in New Zealand, the son of an immigrant farmworker from Scotland and an immigrant schoolteacher from England. Even as a child he showed exceptional academic ability; at 15 he won a scholarship to attend the prestigious Nelson Collegiate School, scoring the highest marks ever recorded on their entrance exams. By 1893 he had obtained a BA and an MA with first-class honours from Christchurch College, Canterbury, and the following year he was the only New Zealander awarded an '1851 Great Exhibition' scholarship to study at Cambridge University, England.

In Cambridge he came under the influence of J J Thomson, who would in 1897 discover the electron. After doing initial work on a sensitive detector for electromagnetic radiation, Rutherford decided to switch his research to the phenomenon of radioactivity, which Henri Becquerel had accidentally discovered in 1896. Rutherford soon found that radioactive emission was more complex than had been initially thought, and was able to identify 'alpha' and 'beta' emission. A few years later, he gave the name 'gamma emission' to the third type of radioactive decay. After four years in Cambridge, Rutherford was offered a professorship at McGill University in Montreal,

Canada, and whilst there he discovered the phenomenon of radioactive half-life. His work on radioactivity would see him win the 1908 Nobel Prize for chemistry 'for his investigations into the disintegration of the elements, and the chemistry of radioactive substances'.

The year 1908 also saw Rutherford returning to England, to take up a professorship at Manchester University. It was there that he made possibly his most famous discovery when, in 1911, working with Hans Geiger and Ernest Marsden he fired alpha particles at gold foil. Entirely unexpectedly, some of the particles bounced back, indicating that the atom has a very dense and massive nucleus. This teamwork was typical of the gregarious Rutherford, who helped transform physics from what had been a primarily solo activity.

In 1917 he bombarded nitrogen gas with alpha particles and found that the gas changed to oxygen – the first time an element had been artificially changed into another one, in effect fulfilling the dreams of the alchemists, though in a way that they could never have conceived. Two years later he was offered the directorship of the Cavendish Laboratories, taking over from Thomson. He became an elder statesman of British science, becoming a peer in 1931 and dying peacefully in Cambridge in 1937.

MACHINES

the 30-second theory

Where would we be without machines? They plough our fields, wash our clothes, transport us around and store our information. They are, in short, devices that perform an action – preferably a useful one. In general, a machine does its job by transforming energy. A waterwheel converts the kinetic energy of flowing water into, say, mechanical energy for grinding corn; a pump might transform mechanical or electrical energy into potential energy of water pumped uphill. The simplest machines might merely transmit force: a wedge or a lever. The most sophisticated have pretensions of intelligence: computer scientists speak of 'machine learning', whereby computers display inductive learning from data that might enable them to make predictions or decisions. Most early machines had moving parts, but in electronic machines the movements tend to be confined to those of electrons, carrying signals and information encoded in electrical currents. No machine is perfectly efficient, using all of the energy it receives to do useful work. Some is inevitably lost as heat, a requirement of the second law of thermodynamics. Thermodynamics was in itself a necessary theoretical development of the age of mechanization – an era accompanied by an increasing tendency to understand life and the human body via analogies with the workings of machinery.

RELATED TOPICS

See also
FORCE & ACCELERATION
page 78

HEAT ENGINES
page 140

SECOND LAW OF
THERMODYNAMICS
page 150

3-SECOND BIOGRAPHIES
ARCHIMEDES
C.287-212 BCE
Greek engineer who described several machines

JULIEN OFFRAY
DE LA METTRIE
1709–51
French physician who saw humans as machines

NORBERT WIENER
1894–1964
American mathematician who developed cybernetics

30-SECOND TEXT
Philip Ball

From ancient waterwheels to modern computer processors, machines transform energy to do work.

3-SECOND THRASH
Machines are devices that use energy to undertake tasks, often involving the conversion of the energy to different forms.

3-MINUTE THOUGHT
Not all machines are human inventions. Biologists, for example, now commonly speak of the living cell as a collection of molecular machines: molecules and structures such as proteins that carry out tasks essential for life, sometimes analogous to the processes effected by large artificial machines, such as mechanical motion (linear and rotary), sensing of signals and even logic processing. Nanotechnologists hope to learn from these natural examples in order to make synthetic nanoscale machinery.

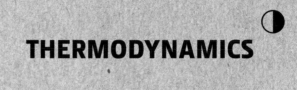

THERMODYNAMICS

absolute zero Temperature combines a measure of the energy of atoms or molecules in a substance and the potential energy of electrons in atoms, which can be raised to higher energy levels (orbits) by absorbing energy. Absolute zero is the temperature at which the atoms in a substance have no kinetic energy and all the atoms are at the lowest energy level, with their electrons in the lowest possible orbits. The temperature of absolute zero is -273.15°C (-459.67°F).

first law of thermodynamics The first law states that energy is conserved in a closed system (one that has no ability to interact with the surrounding universe). Energy can be converted from one form to another, but the total amount in the system remains the same. If the system is not closed, energy changes to match the work done on or by the system, and to cover the heat that flows in or out.

kinetic energy The energy of an object due to its motion. The energy is proportional to the mass of the object and to the square of its velocity. Doubling the velocity quadruples the kinetic energy.

low energy states An atom has energy both from its movement (kinetic energy) and from the position of the electrons within the atom (potential energy). By absorbing energy, usually in the form of photons of light, an electron can be pushed up to a higher orbit, giving the atom extra potential energy. When an atom is in a low energy state, the electrons are all in their lowest possible orbits and the atom has little movement.

potential energy The energy due to the state of a system – for example, the gravitational energy available when an object is lifted up to a high place and can then be dropped, or the energy that is stored in chemical bonds.

quantum 'Quantum' was first used to describe a packet or particle of light when it was discovered that light sometimes behaved as a collection of discrete objects. The term now refers to all objects small enough to be subject to quantum physics.

quantum fluctuations One of the key principles of quantum theory is uncertainty. This describes pairs of properties such as position and momentum that cannot both be known exactly for a quantum particle or system. The more accurately you know one, the less accurate you can be about the other. Another pairing is energy and time. If you look at a quantum system in a very short time frame – so time has to be known very precisely – its energy level can vary considerably. This means that you can't have an atom or object in which all atoms are stopped and at their minimum potential energy levels, because over small periods of time, the energy will vary considerably: these are quantum fluctuations.

second law of thermodynamics In a closed system (one that has no ability to interact with the surrounding universe), heat moves from a hotter to a colder place. Another way of looking at the second law is in terms of entropy (the degree of disorder in a system). In a closed system, the entropy remains the same or increases. It is possible for entropy to decrease by random chance, because this is a statistical law, but the bigger the decrease, the more unlikely it is to happen. If energy can flow into a system, then entropy can decrease

thermodynamics Literally the movement of heat, thermodynamics was a product of the steam age, developed to understand the workings of steam engines. It is primarily concerned with the conservation and flow of energy from place to place in the form of heat.

third law of thermodynamics It is not possible to reach absolute zero in a finite series of steps: in effect, absolute zero is unattainable.

zeroth law of thermodynamics If two objects are in contact so that heat can flow between them, and are in equilibrium (at the same temperature), there will be no net heat flow between them.

HEAT

the 30-second theory

'Feel the heat': the phrase

reflects our intuitive understanding of heat as a form of energy from which notions of hotness derive. But it's a subtle idea. Although it is common to speak of an object's heat content – as if heat were some sort of fluid, as it was once thought to be – in fact heat is strictly speaking energy in motion, being transferred from one body (the hotter one) to another (the cooler). An object feels hot because it transfers energy, as heat, to our fingertips. This heat energy resides in the movements of the atoms and molecules that make up a substance: the hotter the substance, the more vigorously its atoms vibrate, tumble and whizz from place to place. Heat energy may be transferred from one body to another either by the direct contact and collision of their atoms – for example, as it is conducted along a metal rod – or by the emission of electromagnetic radiation through space, as when sunlight (both within and outside the visible spectrum) warms the Earth. In general, heat transfer between two bodies causes a change in their temperature. But it is possible for heat to be transferred without this; for example, when ice at freezing point melts to water at the same temperature. This is called latent heat.

RELATED TOPICS
See also
KINETIC ENERGY
page 122

TEMPERATURE
page 142

SECOND LAW OF
THERMODYNAMICS
page 150

3-SECOND BIOGRAPHIES
JOHN TYNDALL
1820–93
Irish physicist who helped to explain heat as the motion of atoms

HERMANN VON HELMHOLTZ
1821–94
German scientist who explained heat flow as microscopic mechanical motions

30-SECOND TEXT
Philip Ball

3-SECOND THRASH
Heat is a transfer of energy between two bodies. It arises from the motions of their constituent particles.

3-MINUTE THOUGHT
Attempts to understand heat radiation from warm bodies – a long-standing problem in classical physics – led to the inception of quantum theory at the end of the 19th century. The spectrum of radiation could only be understood on the assumption that the vibrations of the body's atoms were quantized – able to adopt some frequencies, but not others. This quantization of vibrations was later generalized to all kinds of energy.

Electromagnetic radiation transfers heat energy through space from the Sun to the Earth.

HEAT ENGINES

the 30-second theory

3-SECOND THRASH
A heat engine converts heat into work, using the flow of heat from hot to cold to do something useful.

3-MINUTE THOUGHT
Heat engines can be run in reverse: an energy source such as electricity can be used to generate heat or move heat in a direction in which it would not otherwise flow, so that work creates a difference in temperature. This is how a refrigerator works: it is a so-called heat pump.

The Industrial Revolution was driven by heat engines. While mechanical power had long been obtained from windmills and water wheels, the invention of the steam engine offered a way to turn heat from burning fuel into mechanical movement through the action of a steam-driven piston. Devices like this (called heat engines) capture some of the energy released as heat and flowing from hot to cold and use it to do useful work. Just as the potential energy of falling water is converted to the kinetic energy of a rotating water wheel, so the movement of heat in a heat engine may lift weights, turn turbines or push a vehicle forwards. For transport, the steam engine was largely replaced by the internal combustion engine; both make use of the fact that a gas expands when it gets hotter. The motion of hot gas also provides the basis for the steam turbine, in which the impact of steam on the turbine blades drives it into rotary motion that can be used to generate electricity. Other heat engines may convert heat directly into electricity without the intermediate mechanical motions: this is how thermoelectric generators work. Heat engines are described by thermodynamic cycles, which link the transfers of heat and work as the temperatures and pressures within the engine vary.

RELATED TOPICS
See also
WORK & ENERGY
page 118

MACHINES
page 130

HEAT
page 138

3-SECOND BIOGRAPHIES
THOMAS NEWCOMEN
1664–1729
English inventor of the first true steam engine

ROBERT STIRLING
1790–1878
Scottish inventor who devised a heat engine using compression and expansion of air

NICOLAS LÉONARD SADI CARNOT
1796–1832
French engineer who launched the discipline of thermodynamics

30-SECOND TEXT
Philip Ball

Steam engine, steam turbine, internal combustion engine – all driven by the expansion of heated gas.

TEMPERATURE

the 30-second theory

Temperature is an everyday concept that is perhaps more complicated than it first appears. Crudely put, it is a measure of the amount of heat in a substance. But some substances can absorb heat more readily than others, so it takes more heat input to raise their temperature. More rigorously expressed, temperature describes how much a certain input of heat adds to the different available configurations of the particles that make up a substance – a quantity related to the material's so-called entropy. Despite these subtleties, temperature is not a hard concept to grasp, because it is generally quite easy to measure, and because it fits so neatly with our tactile sense of hot and cold: hot things have a high temperature. What's more, temperature offers a clear criterion for how heat flows between objects: always from the hotter to the colder. Temperature is traditionally measured using the Celsius and Fahrenheit scales, but physicists prefer the Kelvin scale, because its zero occurs at the lowest possible temperature: absolute zero (-273.15°C or -459.67°F), corresponding to zero heat content. This is impossible to attain in the real world, but scientists have cooled materials down to less than one-billionth of a degree Kelvin (a nanokelvin).

3-SECOND THRASH

Temperature is a measure of hotness – or, more precisely, of how quickly the input of heat changes an object's entropy.

3-MINUTE THOUGHT

Temperatures can be negative even on the Kelvin scale. But this doesn't mean that such objects are colder than absolute zero. They may be so hot that their entropy is almost 'saturated', and adding more heat energy decreases rather than increases it. Some negative-temperature systems are out of thermal equilibrium, so that there are more high-energy than low-energy states: lasers are one example.

RELATED TOPICS
See also
KINETIC ENERGY
page 122

HEAT
page 138

ZEROTH LAW OF
THERMODYNAMICS
page 144

3-SECOND BIOGRAPHIES
ANDERS CELSIUS
1701–44
Swedish astronomer whose temperature scale is based on the freezing and boiling points of water

WILLIAM THOMSON
1824–1907
Northern Irish physicist who first determined the value of absolute zero

HEIKE KAMERLINGH ONNES
1853–1926
Dutch physicist; Nobel Prize for low-temperature physics

30-SECOND TEXT
Philip Ball

You can't go through (or reach) zero on physicists' chosen temperature scale.

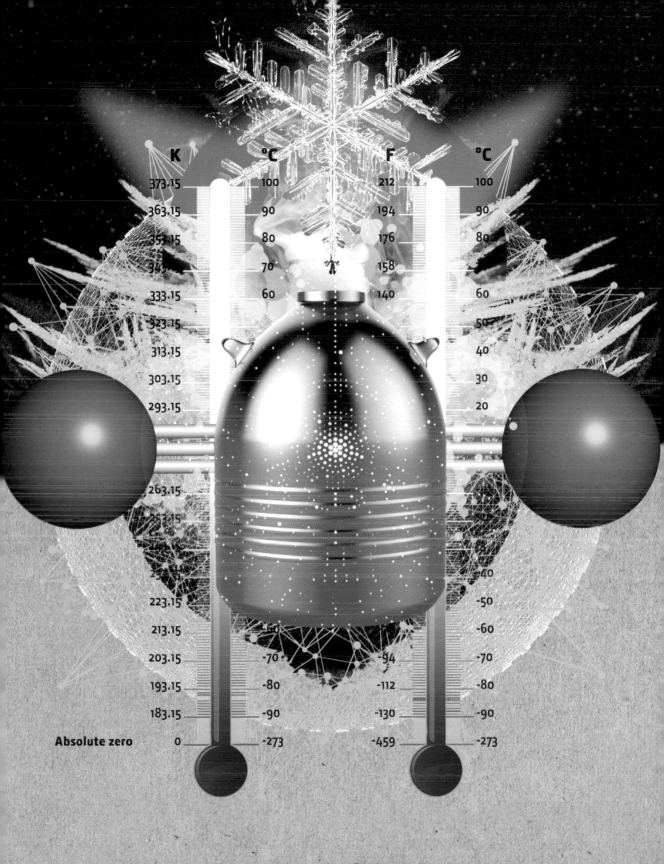

K	°C	F	°C
373.15	100	212	100
363.15	90	194	90
353.15	80	176	80
343.15	70	158	70
333.15	60	140	60
323.15	50		50
313.15			40
303.15			30
293.15	20		20
263.15			
			-40
223.15			-50
213.15	-60		-60
203.15	-70	-94	-70
193.15	-80	-112	-80
183.15	-90	-130	-90
Absolute zero 0	-273	-459	-273

13 June 1831
Born in Edinburgh but afterwards moves to family estate in Galloway

1839
His mother, who has tutored him, dies of cancer

1841
Moves to Edinburgh to live with his aunt and attends the Edinburgh Academy

1846
At the age of 15 he publishes his first scientific paper

1847
Aged 16 he goes to the University of Edinburgh

1850
At 19 he goes to Cambridge University, initially to Peterhouse College but switches to Trinity College after the first term

1854
Graduates from Cambridge, second wrangler in the maths tripos and joint-winner of the Smith Prize

1855
Elected a fellow of Trinity College

1856
His father dies; he is appointed professor of natural philosophy at Marischal College, Aberdeen

1858
Marries Katherine Dewar, daughter of the principal of Marischal College

1860
Made redundant due to a merger between Marischal College and King's College, Aberdeen

1860
Appointed professor of natural philosophy at King's College, London

1861
Produces the world's first colour photograph – of some tartan

1865
Resigns his position at King's College and returns to live on the family estate in Galloway

1871
Appointed professor of experimental physics at Cambridge University and the first director of the Cavendish Laboratories

1873
Publishes his book on electromagnetism *A Treatise on Electricity and Magnetism*

5 November 1879
Dies in Cambridge, and is buried in the church next to his family estate

JAMES CLERK MAXWELL

James Clerk Maxwell is best known for his work on establishing the mathematical relationships that describe electromagnetism; we now call them 'Maxwell's equations'. However, his research ranged much more widely than this. He made important contributions to colour theory, without which we would not have modern-day colour displays in TVs, computers and mobile devices. He also established the speed distribution of molecules in gases (now called the 'Maxwell-Boltzmann distribution') and showed mathematically that Saturn's rings could not be solid rings. He produced the first colour photograph and, in a book on the theory of heat, introduced 'Maxwell's demon', which helped spark the creation of information theory.

Maxwell's parents came from landed gentry and had an estate in Galloway. After coming to Edinburgh for his birth, they moved back to the family estate where Maxwell led an idyllic childhood playing with the local children and being privately tutored by his mother. Tragically, when he was only nine his mother died of cancer, and soon after Maxwell was sent to Edinburgh to live with his aunt and attend the prestigious Edinburgh Academy. He was academically gifted, publishing his first scientific paper on elliptical curves at the age of only 15. At 16, he went to Edinburgh University to study, with the intention of following his father into the legal profession.

At 19, Maxwell left Edinburgh to go to Cambridge University, and four years later he came second in the rigorous maths tripos exam, giving him the title of 'second wrangler'. He did even better in the competition for the Smith Prize, coming joint first. The following year he was made a fellow of Trinity, but at the age of only 24 he was offered a position as professor of natural philosophy at Marischal College in Aberdeen. However, after only four years in this position he was made redundant by the merger of Marischal College and King's College, Aberdeen, to form the new University of Aberdeen. Undeterred, he quickly found another position, as professor of natural philosophy at King's College, London. He would stay in this position for five years, but in 1865 he resigned because he felt that administrative and teaching duties were taking too much time away from his research. Being in the lucky position of being independently wealthy, he returned to live on the family estate in Galloway and do his research in his own time there. Six years later he was tempted back into university life when he was offered the position of the first professor of experimental physics at Cambridge University, and charged with establishing what would become the Cavendish Laboratories. He died in Cambridge in 1879.

Rhodri Evans

ZEROTH LAW OF THERMODYNAMICS

the 30-second theory

RELATED TOPICS

See also
KINETIC ENERGY
page 122

HEAT
page 138

TEMPERATURE
page 142

FIRST LAW OF
THERMODYNAMICS
page 148

3-SECOND BIOGRAPHIES
JAMES CLERK MAXWELL
1831–79
Scottish physicist thought to
have stated a variant of the
zeroth law

CONSTANTIN CARATHÉODORY
1873–1950
German-born Greek
mathematician

RALPH H. FOWLER
1889–1944
English physicist, said to have
devised the title 'zeroth law'

30-SECOND TEXT
Brian Clegg

3-SECOND THRASH
The zeroth law says
that two bodies are in
equilibrium if, despite being
in contact, there is no net
flow of energy from one
body to the other.

3-MINUTE THOUGHT
The zeroth law governs
thermometers. For a
thermometer to work it
has to come into a thermal
equilibrium with the
substance on which it is
making a measurement.
This is why we have to
wait for a traditional
thermometer to reach
the correct value as that
equilibrium is established.
Once there is no net flow
of heat between the
thermometer and the
substance, it will be
showing the appropriate
temperature, as long as it
is correctly calibrated.

What do you do when you
already have a first and second law of
thermodynamics, but decide that there is
something more fundamental required?
Recognizing that '1' is an arbitrary start, the
physics community decided to go for a zeroth
law. The zeroth law is like an axiom in
mathematics. It grounds the other laws by
providing a definition of equilibrium. The law
says that if two objects are in heat equilibrium,
although it is possible for heat to flow from one
to the other, it doesn't. This means that if two
objects with the same temperature are
touching, neither will have an impact on the
temperature of the other. This doesn't mean
that nothing is happening. In practice, energy
will be constantly flowing backwards and
forwards between the two objects as collisions
between atoms or molecules transfer energy
from one body to the other. But according to
the zeroth law, the net flow of energy between
the two is zero. As a direct result of this, we can
see that if A is in equilibrium with B, and C is in
equilibrium with B, then A and C are also in
equilibrium with each other, which is the usual
formulation of the law.

*Three-way equilibrium
– A with B, B with C
and C with A.*

FIRST LAW OF THERMODYNAMICS

the 30-second theory

3-SECOND THRASH

The first law states that energy is always conserved in a system isolated from its surroundings: it may be transformed, but cannot be created or destroyed.

3-MINUTE THOUGHT

The first law rules out the possibility of perpetual motion machines, since these are supposedly able to continue doing work without some constant input of energy to drive them – in effect, to provide something for nothing. This has never stopped people from trying to invent such machines, although the US Patent and Trademark Office has a rule of refusing to grant patents for them.

Energy is constantly being

changed into other forms. The Sun converts nuclear energy to heat and light; in our bodies chemical energy is used to generate movement (kinetic energy), heat, new chemical compounds and electrical impulses in nerves. But all this transformation of energy maintains strict bookkeeping: not a bit of energy goes unaccounted for. That this must be so is what the first law of thermodynamics states: energy is conserved. It can be changed from one form into another, but in a system fully isolated from its surroundings, none is ever gained or lost. In this case, the total energy of the universe is fixed. The first law supplies the basis for understanding the flow of energy in engines and machines. As originally stated by Rudolf Clausius in the mid-19th century it implies that if you want a heat engine – such as a steam engine or internal combustion engine – to do work (like lifting a weight or pumping water) then you have to supply it with heat. To do more work you need more heat: you have to keep supplying fuel. The first law provides the basis of all thermodynamic theory. While it was proposed on simply empirical grounds – experiments seemed to show that energy was conserved, once all forms of it were taken into account – it is now generally considered to be inviolable.

RELATED TOPICS

See also
HEAT
page 138

HEAT ENGINES
page 140

SECOND LAW OF THERMODYNAMICS
page 150

3-SECOND BIOGRAPHIES

WILLIAM RANKINE
1820–72
Scottish engineer, with Rudolf Clausius the first to state the conservation of energy

RUDOLF CLAUSIUS
1822–88
German physicist who formulated a version of the first law

MAX BORN
1882–1970
German physicist who reformulated the first law in precise mathematical terms

30-SECOND TEXT

Philip Ball

Clausius understood that heat engines need heat (supplied by burning fuel) to do work.

SECOND LAW OF THERMODYNAMICS

the 30-second theory

This is the most interesting of the laws of thermodynamics, because it tells us how stuff happens. Specifically, it stipulates that in all natural processes the total entropy of the universe increases. (Strictly speaking that applies to change in any isolated system that can't exchange heat with the outside – the universe is presumed to be such a system.) Entropy is a measure of the disorder in the system: it measures the number of different ways in which the components of the system can be arranged, and so is larger for less orderly systems. The second law is simply a question of probability: high-entropy states, being more numerous, are more likely to arise from processes of change than are low-entropy states. This becomes less true as the systems get smaller and the options are more constrained, which is why the second law is then less prescriptive about what will happen. Some scientists think that it is better to express the second law in terms of energy dispersal or dissipation: energy always tends to spread out, so that for example heat always flows from hot to cold. When change produces order and organization – the growth of a snowflake or a living organism – the second law is preserved because the process also generates heat that creates compensating disorder in the surroundings.

3-SECOND THRASH

The second law states that the total entropy in any isolated system always increases during a process of change, because that is overwhelmingly most probable.

3-MINUTE THOUGHT

Some researchers believe that the second law defines the arrow of time: why it seems only to go forwards. The basic laws of motion work in either time direction: a movie of two colliding billiard balls makes sense in reverse. But entropy can only increase in one direction – ink drops in water don't unmix, shattered vases don't become whole again. Whether there are more fundamental reasons for time's arrow, and how to explain our perception of its forward flow, are still matters of debate.

RELATED TOPICS

See also
FIRST LAW OF THERMODYNAMICS
page 148

THIRD LAW OF THERMODYNAMICS
page 152

3-SECOND BIOGRAPHIES

RUDOLF CLAUSIUS
1822–88
German physicist who introduced the concept of entropy

LUDWIG BOLTZMANN
1844–1906
Austrian physicist who interpreted the second law in terms of probability

ROLF LANDAUER
1927–99
German-born American physicist who connected the second law to information theory

30-SECOND TEXT
Philip Ball

When an apple decays and falls apart its entropy increases – it becomes more disordered.

THIRD LAW OF THERMODYNAMICS

the 30-second theory

RELATED TOPICS
See also
ATOMS
page 16

THE UNCERTAINTY PRINCIPLE
page 64

TEMPERATURE
page 142

SECOND LAW OF
THERMODYNAMICS
page 150

3-SECOND BIOGRAPHIES
WALTHER NERNST
1864–1941
German physicist who
developed the third law
of thermodynamics

WOLFGANG KETTERLE
1957–
German physicist who
demonstrated apparent
negative absolute
temperatures in a
magnetic system

30-SECOND TEXT
Brian Clegg

3-SECOND THRASH
The third law says that at
absolute zero the entropy
of a body is zero and this
can't be achieved in a finite
number of steps.

3-MINUTE THOUGHT
Although you can't
get anything through
absolute zero, it is
theoretically possible to
have something the other
side. Temperature is
statistical measure of the
distribution of kinetic
energy of particles.
Temperature goes up as the
distribution spreads. But if
most particles have similar
very high energies, the
result is a flip to a negative
absolute temperature. The
particles approach absolute
zero from beneath as
energies increase and
entropy reduces again.
Some physicists argue this
is not a true negative
temperature, but most
accept it.

Where the earlier laws of
thermodynamics are fundamentally products
of the steam age, the third law is more a matter
for the quantum age. The law says that it is not
possible to get anything down to absolute zero,
which is 0 K or -273.15°C (-459.67°F), in a finite
series of steps. Temperature is a measure of the
energy of the atoms or molecules in a substance.
At absolute zero they would have reached the
lowest possible level, both in terms of kinetic
energy and the energy levels of electrons within
the atom. But in practice, because of the atom's
quantum nature, which means that energy levels
will naturally fluctuate, it is impossible to reach
absolute zero. Another way of looking at the
third law is that temperature approaches
absolute zero as the entropy of the object
decreases, because the atoms move less and
can occupy fewer and fewer energy states, so
the number of states the whole body can be in
– a definition of its entropy – gets lower and
lower. Even without an awareness of quantum
fluctuations, the step-by-step nature of reducing
the number of possible states makes it
mathematically impossible to achieve the final
step at absolute zero.

*Physicists have got to
within one-billionth K
of absolute zero.*

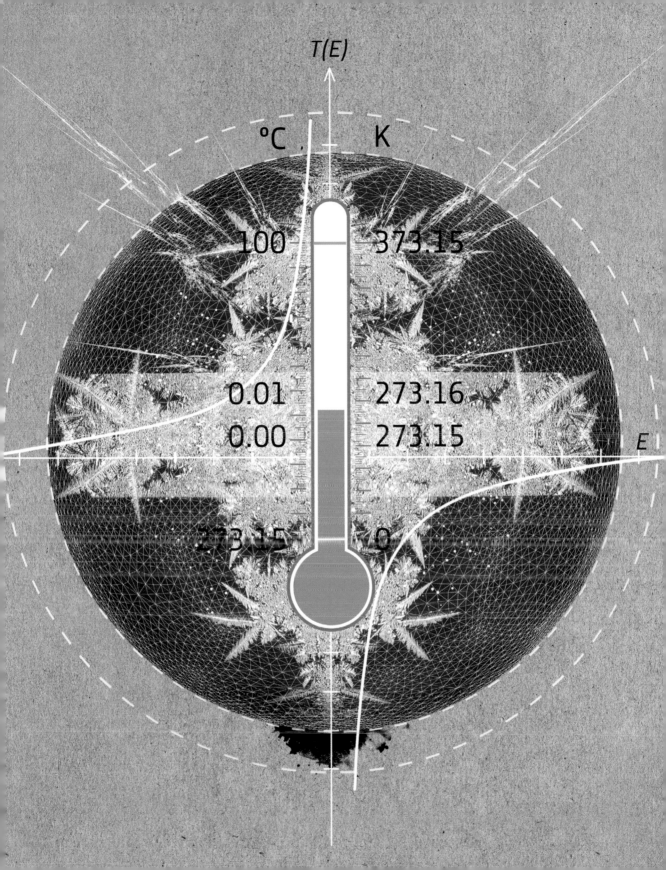

RESOURCES

BOOKS

Antimatter
Frank Close
(Oxford University Press, 2010)

Beam: The Race to Make the Laser
Jeff Hecht
(Oxford University Press, 2010)

Before the Big Bang
Brian Clegg
(Saint Martins Griffin, 2011)

Black Holes, Wormholes and Time Machines
Jim Al-Khalili
(Taylor & Francis, 2012)

A Brief History of Infinity: The Quest to Think the Unthinkable
Brian Clegg
(Robinson Publishing, 2003)

A Brief History of Time
Stephen Hawking
(Bantam, 2011)

Build Your Own Time Machine: The Real Science of Time Travel
Brian Clegg
(Gerald Duckworth & Co., 2013)

Compendium of Theoretical Physics
Armin Wachter and Henning Hoeber
(Springer, 2005)

Dice World: Science and Life in a Random Universe
Brian Clegg
(Icon Books, 2013)

Erwin Schödinger and the Quantum Revolution
John Gribbin
(Black Swan, 2013)

The Fifth Essence
Lawrence Krauss
(Vintage, 1990)

The Infinity Puzzle
Frank Close
(Oxford University Press, 2011)

In Search of Schrödinger's Cat
John Gribbin
(Black Swan, 1985)

The God Effect: Quantum Entanglement, Science's Strangest Phenomenon
Brian Clegg
(Saint Martins Griffin, 2009)

*Paradox: The Nine Greatest
Enigmas in Physics*
Jim Al-Khalili
(Black Swan, 2013)

Particle Physics: An Introduction
Frank Close
(Oxford University Press, 2004)

*QED: The Strange Theory of Light
and Matter*
Richard Feynman
(Penguin, 1990)

The Road to Reality
Roger Penrose
(Vintage, 2005)

Why Does E=MC2?
Brian Cox and Jeff Forshaw
(De Capo, 2010)

WEBSITES
Eric Weisstein's World of Physics
scienceworld.wolfram.com/physics/

Frequently Asked Questions in Physics
math.ucr.edu/home/baez/physics/

JOURNALS/ARTICLES
Do tachyons exist?
math.ucr.edu/home/baez/physics/
ParticleAndNuclear/tachyons.html

Quantum Entanglement and Information,
Stanford Encyclopedia of Philosophy
plato.stanford.edu/entries/qt-entangle/

Testing the Multiverse, article on FQXI
website by Miriam Frankel
fqxi.org/community/articles/display/155

Parallel Universes by Max Tegmark,
Scientific American, 2003
space.mit.edu/home/tegmark/PDF/
multiverse_sciam.pdf

*Faster than the speed of light? We'll need
to be patient by Jim Al-Khalili*
www.guardian.co.uk/commentisfree/2011/
nov/23/faster-speed-of-light-boxers

NOTES ON CONTRIBUTORS

Philip Ball is a freelance writer, and was an editor for *Nature* for more than 20 years. Trained as a chemist at the University of Oxford, and as a physicist at the University of Bristol, he writes regularly in the scientific and popular media, and has authored books including *H2O: A Biography of Water, Bright Earth: Art and the Invention of Colour, The Music Instinct: How Music Works and Why We Can't Do Without It, and Curiosity: How Science Became Interested in Everything*. His book *Critical Mass: How One Thing Leads to Another* won the 2005 Aventis Prize for Science Books. He has been awarded the American Chemical Society's Grady–Stack Award for interpreting chemistry to the public, and was the inaugural recipient of the Lagrange Prize for communicating complex science.

Brian Clegg read Natural Sciences, focusing on experimental physics, at the University of Cambridge. After developing hi-tech solutions for British Airways and working with creativity guru Edward de Bono, he formed a creative consultancy advising clients ranging from the BBC to the Met Office. He has written for *Nature*, the *Times*, and the *Wall Street Journal* and has lectured at Oxford and Cambridge universities and the Royal Institution. He is editor of the book review site www.popularscience.co.uk, and his own published titles include *A Brief History of Infinity* and *How to Build a Time Machine*.

Leon Clifford is a writer and a consultant and a business director whose speciality is simplifying complexity. Leon has a BSc in physics-with-astrophysics and is a member of the Association of British Science Writers. He worked for many years as a journalist covering science, technology and business issues with articles appearing in numerous publications including *Electronics Weekly, Wireless World, Computer Weekly, New Scientist*, and the *Daily Telegraph*. Leon is interested in all aspects of physics – particularly climate science, astrophysics and particle physics. He is managing director of science communications business Green Ink which focuses on the challenges of communicating science for development.

Frank Close, OBE, is Professor of Physics at the University of Oxford and Fellow of Exeter College, Oxford. He was formerly Head of the Theoretical Physics Division, at Rutherford Appleton Laboratory and Head of Communications and Public Education at CERN. His research is into the quark and gluon structure of nuclear particles, where he has published more than 200 papers in the peer-reviewed literature. He is a Fellow of the American Physical Society, and of the British Institute of Physics, and won the society's Kelvin Medal in 1996 for his outstanding contributions to the public understanding of physics. He is the author of many books including *Neutrino* – short-listed for the Galileo Prize in 2013 – the best selling *Lucifer's Legacy: The Meaning of Asymmetry*, and, most recent, *The Infinity Puzzle*

Rhodri Evans studies and researches in extra-galactic astronomy. Rhodri has, for over 16 years, been involved in airborne astronomy, and is a key part of the team building the facility far-infrared camera for SOFIA. He is also involved in research into star-formation and cosmology, and is a regular contributor to television, radio and public lectures. Rhodri runs the blog www.thecuriousastronomer. wordpress.com.

Andrew May is a technical consultant and freelance writer on subjects ranging from astronomy and quantum physics to defence analysis and military technology. After reading Natural Sciences at the University of Cambridge in the 1970s, he went on to gain a PhD in Astrophysics from the University of Manchester. Since then he has accumulated more than 30 years' worth of diverse experience in academia, the scientific civil service and private industry.

INDEX

INDEX

ACKNOWLEDGEMENTS

PICTURE CREDITS
The publisher would like to thank the following individuals and organizations for their kind permission to reproduce the images in this book. Every effort has been made to acknowledge the pictures; however, we apologize if there are any unintentional omissions.

Alamy: 72.
Library of Congress Prints and Photographs Division Washington, D.C.: 112, 132.
Shutterstock: 30, 52, 92, 152.